DIFFRACTION ANALYSIS OF DEFORMED METALS

THEORY, METHODS, PROGRAMS

FAINA F. SATDAROVA

THE THIRD REVISED AND ENLARGED EDITION

ACADEMUS Publishing

Academus Publishing
2019

Academus Publishing, Inc.

1999 S, Bascom Avenue, Suite 700 Campbell CA 95008
Website: www.academuspublishing.com
E-mail: info@academuspub.com

The right of Faina F. Satdarova,
is identified as author of this work.

The third revised and enlarged edition

Editor of the English translation:
Faina F. Satdarova, Moscow, Russia.

ISBN 10: 1 4946 0016 1
ISBN 13: 978 1 4946 0016 7
DOI 10.31519/monography_1598

General analysis of the distribution of crystals orientation and dislocation density in the polycrystalline system is presented.

Recovered information in diffraction of X-rays adopting is new to structure states of polycrystal. Shear phase transformations in metals – at the macroscopic and microscopic levels – become a clear process.

Visualizing the advances is produced by program included in package delivered. Mathematical models developing, experimental design, optimal statistical estimation, simulation the system under study and evolution process on loading serves as instrumentation.

To reduce advanced methods to research and studies problemoriented software will promote when installed. Automation programs passed a testing in the National University of Science and Technology "MISIS" (The Russian Federation, Moscow).

You score an advantage in theoretical and experimental research in the field of physics of metals.

DEDICATION

To Petrunenkov Alexander Alexandrovich
with great gratitude

FOREWORD

Development of optimal on the accuracy methods of research of the structure of polycrystalline systems began in Moscow Institute of steel and alloys, and continued (after an enforced break) as a private scientific activity.

Best analysis of diffraction observations on the basis of established theoretical models brought new knowledge about the plastic deformation structures.

The third edition was supplemented by the method of determining the long-range order in the dislocation structure of crystals, which in practice gave important results for understanding the kinetics of martensitic transformation during the hardening of steel (Ch. 6–7 and 10).

A fundamentally new idea of shear transformations in metals originated: first of a macroscopic level (textural transformation), and now of a microscopic level (martensitic transformation).

The methodology embodied in the high-tech software products, which are a tool to automate the research of deformed metal with crystals of cubic symmetry, constituting the main class of materials.

X-ray measuring test specimens were performed by Kozlov Dmitry Alexandrovich, with his exceptional integrity.

F.F. Satdarova

INTRODUCTION

Macroscopic structural states of a deformed polycrystalline system are determined by the degree of ordering of the orientations of crystals. Each macroscopic state has some distribution of microstates, characterized by the parameters of random system of dislocations in inhomogeneously deformed crystals [72].

Theoretical analysis of the kinetics of the crystal orientation distribution suggests the following conceptual model of structural transformation of a polycrystalline system [76]:

- In the area of non-critical deformations, nonequilibrium fluctuations of crystal orientations evolve, and a short-range orientational order is formed (continuous change of metastable states).
- When deformation reaches the critical value, the collective of crystals undergo instantaneous consistent rotation, a long-range orientational order arises, and a structure with special symmetry is formed (abrupt transition into a qualitatively new stable state).

At the transformation instant, the fluctuation of crystal orientations in which the plastic strain rate increases more rapidly is sharply localized in the orientation space; the other fluctuations "frozen". This is evidenced by the formation of shear bands with macroscopic orientational order and a sharp increase in the dislocation density in closely oriented crystals [52, 74].

Self-organization of strongly non-equilibrium polycrystalline system is hypothesized as natural way of forming the crystallographic texture. In pursuit of the overall insensitively order a dislocation fluctuation is born, creating intra-crystalline disorder. It has developed a holistic picture of the structure states of deformed metals.

The method of achieving the objective of the study is subject to unified principle: high quality theoretical model, the best experiment to estimate of the model, maximum possible accuracy of the physical parameters by model.

PART I
CRYSTALLOGRAPHIC TEXTURE OF DEFORMED METALS

Mathematical model for the emergence of scientific direction predicts that in the decision the problem, which not be solved; large scientific community is involved [9]. The confirmation became "a recovering the distribution function of orientations from the pole figures" [7]. The diversity of mathematical methods created to processing traditional texture measurements is presented into qualified overview [89].

The pole figures, where clearly expressed the symmetry of the crystallographic texture, the density probability of the orientations of crystals (as understood by the distribution function of orientations) appeared in a "ghost" mode. Orientation probability distribution is not obvious substituted by the function reproducing the observed fluctuations of orientations. Computational methods are powerless to extract information, which is not in the experiment.

To distribution of random orientations in polycrystalline material, statistical estimation method is objectively predefined [64, 71]:

1. Theoretical formulation of probability distribution in accordance with the physical nature of the texture.

2. Experiment seeks to acquiring information on the distribution of the orientations of crystals.

3. Reliable estimates of the distribution parameters by the experimental data.

Texture function that represents the essence of the object has true theoretical and practical usefulness to explore the properties of deformed metals.

CHAPTER 1
THE ORETICAL PROBABILITY DISTRIBUTION OF THE CRYSTAL ORIENTATIONS. TEXTURE FUNCTION

In the theoretical texture functions general mathematical representation is driven to conformity with physical limitations and experience. Physical limitations follow from symmetry of object. Experience revealed that is the most significant in the object.

§ 1.1. Invariance of the Texture Function Relative to Symmetry Transformations

Probability distribution of the crystal orientations has the Fourier representation in the space of generalized spherical functions consistent with representation of continuous groups of three-dimensional rotations [99].

1. The space of representation of the texture function. Crystal orientation by rotating the lattice basis is described in the coordinate system (X, Y, Z). Texture function $f(\mathbf{g})$, where $\mathbf{g}$ is the vector that defines the rotation in three-dimensional space, is considered to be specified on the group of rotations $\mathbf{G}$. What is implied it is an abstract infinite group with constantly changing of parameters of elements of $\mathbf{g}$, and therefore the group is called continuous.

When the matrix representing rotation built in normal space with a basis of three unit vectors $(\mathbf{u}_1, \mathbf{u}_2, \mathbf{u}_3)$, group parameters are projections of rotation axis on the coordinate axes multiplied by a rotation angle. When extending matrix representation the group of rotations over to a space of functions as parameters choose Euler angles. When Eulerian coordinates any rotation is expressed as product of three simple rotations around the coordinate axes $\mathbf{u}_1$, $\mathbf{u}_2$, $\mathbf{u}_3$ [37].

Introduced different designations Euler angles:

Korn [37]	*Vilenkin* [100]	*Viglin* [99]	*Roe* [56]
α	$\varphi = \alpha + \pi/2$	$\varphi_2 = \alpha$	$\psi = \alpha$
β	$\vartheta = \beta$	$\vartheta = \beta$	$\vartheta = \beta$
γ	$\psi = \gamma - \pi/2$	$\varphi_1 = \gamma$	$\varphi = \gamma$

Let us adopt the notation R. Roe [56], who designed the foundations of the harmonic analysis of pole figures. Elements of continuous rotation group will as follow

$$\mathbf{g}(\psi, \vartheta, \varphi) = \mathbf{g}(\psi, 0, 0)\, \mathbf{g}(0, \vartheta, 0)\, \mathbf{g}(0, 0, \varphi)$$

$$(0 \le \psi \le 2\pi), \quad (0 \le \vartheta \le \pi), \quad (0 \le \varphi \le 2\pi).$$

Group of matrices of size $(2l+1)\times(2l+1)$, representing the rotation with Euler angles, is composed of elements of type $T^l_{mn}(\mathbf{g})$, where l is the maximum value of m indexes for abstract basis vectors $\mathbf{u}_m(\mathbf{g})$ of functional space, and n is indexes for new basis vectors after rotation of basis. The number l can take values $(0, {}^1/_2, 1, {}^3/_2, 2, \ldots)$, but half-integer representation is not periodic on 2π range (double-digit) [16].

For integer l the basis vectors of $(2l+1)$-dimensional representation space are orthonormal spherical functions [37]:

$$Y_{lm}(\vartheta,\psi)=\sqrt{\frac{2l+1}{4\pi}\frac{(l-|m|)!}{(l+|m|)!}}\,P_l^{|m|}(\cos\vartheta)e^{-im\psi}.$$

Here, $P_l^m(\cos\vartheta)$ is the attached Legendre functions of degree of l and order of m ($l = 0, 1, 2, \ldots$ ($m = 0, \pm1, \pm2, \ldots, \pm l$)); ϑ, ψ is spherical coordinates. (This designation of the spherical functions is adopted in the field of theoretical physics, although with a changing arrangement of the indexes.)

Basis spherical functions of rotated basis with transformed coordinates have decomposition on the original basis functions. The decomposition coefficients exactly are the matrix elements of representation the group of rotations:

$$Y_{ln}(\vartheta',\psi')=\sum_{m=-l}^{l}T^l_{mn}(\mathbf{g})Y_{lm}(\vartheta,\psi).$$

Matrix elements $T^l_{mn}(\mathbf{g})$ form a space in which they are orthogonal as m, n, and l:

$$(2l+1)\int_{\mathbf{G}}T^l_{mn}(\mathbf{g})^*\,T^{l'}_{m'n'}(\mathbf{g})\,d\mathbf{g}=\delta_{ll'}\,\delta_{mm'}\,\delta_{nn'}$$

($T^l_{mn}(\mathbf{g})^*$ is complex conjugate element) [16].

Considered as a function of continuous rotation $\mathbf{g}(\psi,\vartheta,\varphi)$ parameters, matrix elements $T^l_{mn}(\mathbf{g})$ are the generalized spherical functions.[1] They contain the generalized attached Legendre functions [100]:

$$T^l_{mn}(\mathbf{g})=e^{-im\psi}P_l^{mn}(\cos\vartheta)e^{-in\varphi}.$$

---

[1] A lot of reference information about spherical functions collected in a book Bunge [7].

Private expressions of $P_l^{mn}(t)$, where $t = \cos\vartheta$, are

$$P_l^{m0}(t) = \sqrt{\frac{(l-m)!}{(l+m)!}}\,P_l^m(t),\quad P_l^{0n}(t) = \sqrt{\frac{(l-n)!}{(l+n)!}}\,P_l^n(t).$$

Any function $f(\mathbf{g})$ on group **G**, such that

$$\int_{\mathbf{G}} \left|f(\mathbf{g})\right|^2 d\mathbf{g} < +\infty,$$

can be expanded into converging on average Fourier series by function $T_{mn}^l(\mathbf{g})$ [100]:

$$f(\mathbf{g}) = \sum_{l=0}^{\infty}\sum_{m=-l}^{l}\sum_{n=-l}^{l} W_{lmn} T_{mn}^l(\mathbf{g}),$$

$$W_{lmn} = (2l+1)\int_{\mathbf{G}} f(\mathbf{g}) T_{mn}^l(\mathbf{g})\,d\mathbf{g},$$

$$\int_{\mathbf{G}} \left|f(\mathbf{g})\right|^2 d\mathbf{g} = \sum_{l=0}^{\infty}(2l+1)^{-1}\sum_{m=-l}^{l}\sum_{n=-l}^{l}\left|W_{lmn}\right|^2.$$

Groups of symmetry both of crystal lattice and of sample, having a finite number of elements, are subgroups of continuous rotation group. Requirement of invariance of the texture function $f(\mathbf{g})$ to the symmetry transformations imposes strict limitations on its representations in the chosen basis of generalized spherical functions $T_{mn}^l(\mathbf{g})$.

2. Valid harmonics of the texture function under cubic symmetry of crystals. By definition R. Roe [56] the angles ϑ and ψ are spherical coordinates of the basis vector of crystal lattice in the coordinate system of the sample, φ is angle of rotation crystal around its own basis vector. Crystallographic orientation (*hkl*)[*uvw*] is described by Euler angles (ψ,ϑ,φ):

$$\cos\vartheta = \frac{l}{\sqrt{h^2+k^2+l^2}},\quad \cos(\varphi+\psi) = \frac{u}{\sqrt{u^2+v^2+w^2}}\qquad (h=k=0),$$

$$\cos\varphi = \frac{-h}{\sqrt{h^2+k^2}},\quad \cos\psi = \frac{w}{\sqrt{u^2+v^2+w^2}}\frac{\sqrt{h^2+k^2+l^2}}{\sqrt{h^2+k^2}}\qquad (h,k\neq 0).$$

In the group of cubic symmetry, which refers to point groups (all axis and planes of symmetry contain a fixed point that is any lattice site), are only 24 elements. This is four rotations around the axis ⟨001⟩ on the angle $\pi/2$ (rearranging of indexes h, k and changing their sign), three rotations around the axis ⟨111⟩ on the angle $2\pi/3$ (a cyclic rearranging of indices $(hkl)[uvw]$) and two rotations around the axis ⟨110⟩ on the angle π (changing sign l and w).

Valuable a priori information about the harmonics $f(\mathbf{g})$ is acquired through decomposition of representations of a continuous rotation group as to representations of crystallographic point group [16]. The information obtained by methods of the theory of groups for cubic symmetry of crystals is shown in Table 1.1.

Table 1.1.

Invariants relative to the group of cubic symmetry in the representation of three-dimensional rotation

Number of invariants	*The degree of spherical harmonics*	*Theoretical information on Fourier coefficients of the texture function*
0	$l = 1, 2, 3, 5, 7, 11$	All coefficients W_{lmn} are zero.
1 (a)	$l = 4, 6, 8, 10, 14$	$W_{lmn} = 0$ if $\lvert m\rvert \neq 2k$, $\lvert n\rvert \neq 4k$ $(k = 0, 1, 2, \ldots)$; W_{lmn}, where $n = \pm 4, \pm 8, \ldots$ $(\lvert n\rvert \leq l)$, are linearly related to W_{lm0} $(\lvert m\rvert = 0, 2, 4, \ldots, l)$.
1 (b)	$l = 9, 13, 15, 17, 19, 23$	$W_{lmn} = 0$ if $n = 0$ $(\lvert m\rvert = 0, 2, 4, \ldots)$; W_{lmn}, where $n = \pm 8, \pm 12, \ldots$ $(\lvert n\rvert < l)$, are linearly related to $W_{lm4} = W^*_{lm\bar{4}}$ $(\lvert m\rvert = 0, 2, 4, \ldots)$.
2 (a)	$l = 12, 16, 18, 20, 22, 26$	W_{lmn}, where $n = \pm 8, \pm 12, \ldots$ $(\lvert n\rvert \leq l)$, are linearly related to W_{lm0} и $W_{lm4} = W^*_{lm\bar{4}}$ $(\lvert m\rvert = 0, 2, 4, \ldots, l)$.
2 (b)	$l = 21, 25, 27, 29, 31, 35$	W_{lmn}, where $n = \pm 12, \pm 16, \ldots$ $(\lvert n\rvert < l)$, are linearly related to $W_{lm4} = W^*_{lm\bar{4}}$ and $W_{lm8} = W^*_{lm\bar{8}}$ $(\lvert m\rvert = 0, 2, 4, \ldots)$.

Due to the symmetry of the crystal lattice occurs disappearance of the series coefficients in expansion of $f(\mathbf{g})$. The non-zero Fourier coefficients are W_{lmn} only the even order of m and multiple of four of an order of n. For even m and n relations between the harmonics are simplified:

$$W_{lmn} = W_{l\bar{m}n} = W_{lm\bar{n}} = W_{l\bar{m}\bar{n}} \quad (\bar{m} = -m,\ \bar{n} = -n).$$

Moreover there are relations of the Fourier coefficients W_{lmn} of different orders of n indicated in the Table 1.1. Linear equations relating W_{lmn} with even degree $l \leq 22$ (at $l = 24$ number of invariants reaches three) are given in the article by R. Roe [57].

All Fourier coefficients W_{lmn} odd degrees of l with the order of $n = 0$ turn to zero. For following n they are linearly related. For example,

$$W_{9m8} = -0.64168895\, W_{9m4} \qquad (m = 0, 2, \ldots, 8),$$

$$\left.\begin{aligned} W_{13m8} &= 0.29019050\, W_{13m4} \\ W_{13m12} &= -0.72981613\, W_{13m4} \end{aligned}\right\} \quad (m = 0, 2, \ldots, 12).$$

The proportion of independent spherical harmonics of even and odd degree l in the Fourier representation of $f(\mathbf{g})$ can be seen from Table 1.2.

Table 1.2.

Numbers of independent harmonics of even and odd degree in the Fourier representation of the texture function

Type *	*The limiting degree of harmonics*										
	4	*6*	*8*	*10*	*12*	*14*	*16*	*18*	*20*	*22*	*24*
(a)	3	8	13	19	33	41	59	79	101	125	164
(b)	0	0	0	5	5	12	20	29	39	61	73

* (a) even degree; (b) odd degree.

Any rotation $\mathbf{g}$, mixing the related to T^l_{mn} spherical basis functions Y_{lm} of the same degree l, when decomposition along original basis, is mixing as well Fourier coefficients W_{lmn} with identical l, n:

$$W_{lm'n} = \sum_{m=-l}^{l} T^l_{mm'}(\mathbf{g}) W_{lmn}.$$

The only coefficient – W_{000} is invariant with respect to all rotations of three-dimensional space.

The full group of orthogonal transformation in three-dimensional space, denoted as $\mathbf{O}_3$ except rotations includes inversion, which change the direction of the coordinate axes X, Y, Z. Subsequent rotation, for example, around the Z axis by angle π returns direction the X, Y axis, the reflection in the horizontal plane will remain. At rhombic symmetry of a sample there are three reflection planes perpendicular to the axes of rotation forming the coordinate system (X, Y, Z).

For invariance $f(\mathbf{g})$ with respect to reflection in the horizontal symmetry plane of a sample the Fourier series should consist of symmetric representations

$$\frac{1}{2}\left[T^l_{mn}(\psi,\vartheta,\varphi)+T^l_{mn}(\psi,\pi-\vartheta,\varphi)\right].$$

Toward reflections in the vertical planes of symmetry of a sample $f(\mathbf{g})$ is invariant only when the imaginary component of the Fourier coefficients $W_{lmn}=\frac{1}{2}\left(U_{lmn}+iV_{lmn}\right)$ equal to zero, since $T^l_{mn}(\mathbf{g})$ when reflected in the vertical planes are converted into complex conjugate [16].

Mathematical expression for harmonics of the texture function appears in a statistical model of the probability distribution of crystal orientations.

§ 1.2. Probability Density of the Orientations of Crystals in the Rhombic Texture

Theoretical probability density of orientations is forming by a model of spherical normal dispersion of the crystallographic vectors relative to the directions of ordering really existing for a plane deformed metals [71].

1. Mixed probability distribution of the orientations in polycrystalline system. Mathematical description of the real crystallographic texture is function

$$\left.\begin{aligned} &f(\mathbf{g},\mathbf{B})=\sum_{\nu=1}^{s}\mu_\nu f^{(\nu)}(\mathbf{g},\mathbf{C}_\nu),\\ &\mathbf{B}^{\mathrm{t}}=\left(\mathbf{B}^{\mathrm{t}}_1,\ \ldots,\ \mathbf{B}^{\mathrm{t}}_s\right),\ \ \mathbf{B}^{\mathrm{t}}_\nu=\left[\mu_\nu,\mathbf{C}_\nu\right],\end{aligned}\right\} \qquad (1.1)$$

where μ_ν is weight fraction of ν-th component of the texture, having a density distribution of orientations $f^{(\nu)}(\mathbf{g},\mathbf{C}_\nu)$ with the parameters of $\mathbf{C}_\nu$ (t-superscript denotes transpose).

Each of texture components, the number of which is s, in three-dimensional space of rotations of $\mathbf{G}$ being presented as the set of density maxima of the orientations $\mathbf{E}=\left(\mathbf{e}_1,\ldots,\mathbf{e}_p\right)$ in accord with the symmetry transformations of crystal (p is the repetition factor).

Discrete distribution of ordered crystallographic orientations in $\mathbf{G}$ can be approximated by the density function

$$R(\mathbf{g},\mathbf{E})=\frac{1}{p}\sum_{k=1}^{p}\delta\left(\mathbf{e}_k^{-1}\mathbf{g}\right). \qquad (1.2)$$

Transformation of the δ-function with $\mathbf{g}$ was induced by rotation, which is involved in the elements of group $\mathbf{E}$ [16].

Let us introduce the function $Q(\mathbf{g},\mathbf{K})$, which describes a random dispersion of orientations with respect to the expected by **E**. Parameter **K** is measure of dispersion. Probability density of crystal orientations in a given texture component is determined by the convolution of the functions $Q(\mathbf{g},\mathbf{K})*R(\mathbf{g},\mathbf{E})$.

The form of the functions $Q(\mathbf{g},\mathbf{K})$, $R(\mathbf{g},\mathbf{E})$ is the same for all $f^{(\nu)}(\mathbf{g},\mathbf{C}_\nu)$, only their parameters are changing, which are components of the vector $\mathbf{C}_\nu = [\mathbf{E}_\nu, \mathbf{K}_\nu]^t$ (ν = 1, … , s). Therefore in the following mathematical expressions ν index is dropped.

Let the external coordinate system is built according to the rhombic symmetry of plane deformed metal, and let the most probable crystallographic orientation at this system is (001)[100]. Through choice of the coordinate system in crystal any orientation can be converted to that type.

It is reasonably to suggest that the dispersion of the orientations of crystals combines the deviations of crystallographic planes (001) from plane of the deformation and the deviations of crystallographic directions [100] lying in the plane from the rolling axis. The observed random (*hkl*)[*uvw*] orientation has a rotation $\mathbf{g}(\psi,\vartheta,\varphi)$ [12]:

$$\begin{cases} h = -\sin\vartheta\cos\varphi, & u = \cos\psi\cos\vartheta\cos\varphi - \sin\psi\sin\varphi, \\ k = \sin\vartheta\sin\varphi, & v = -\cos\psi\cos\vartheta\sin\varphi - \sin\psi\cos\varphi, \\ l = \cos\vartheta, & w = \cos\psi\sin\vartheta. \end{cases}$$

When exposed to a large number the dispersion factors existing in reality, the occurrence probability of random (*hkl*) and [*uvw*] near the expected (001) and [100] are theoretically subject to the normal law [19, 6]:

$$Q_{(hkl)\,(001)} \sim e^{\mathrm{K}_1\cos\vartheta}, \quad Q_{[uvw]\,[100]} \sim \delta(\vartheta)\, e^{\mathrm{K}_2(\cos\psi\cos\vartheta\cos\varphi - \sin\psi\sin\varphi)};$$

K_1 and K_2 is probability distribution parameters (similar to the inverse of the variance).

It can suggest that deviations the vectors of a crystal lattice around the ordering directions are independent random variables. Convolution of functions $Q_{(hkl)\,(001)}$ and $Q_{[uvw]\,[100]}$ will give the spherical normal model of the orientations probability density:

$$Q(\mathbf{g},\mathbf{K}) = \left[\left(\frac{\operatorname{sh}\mathrm{K}_1}{\mathrm{K}_1}\right) I_0(\mathrm{K}_2)\right]^{-1} e^{[\mathrm{K}_1\cos\vartheta + \mathrm{K}_2\cos(\varphi+\psi)]}, \qquad (1.3)$$

$$\mathbf{K}=\begin{bmatrix}K_1\\K_2\end{bmatrix};\ \int_{\mathbf{G}} Q(\mathbf{g},\mathbf{K})d\mathbf{g}=1,\ d\mathbf{g}=\frac{1}{8\pi^2}d\varphi\sin\vartheta d\vartheta d\psi;$$

$I_n(x)$ is a modified Bessel function [24].

The introduced function is a statistical model for the distribution of orientations of the crystals, approximately representing the physical reality. The acceptability of model assumptions about the independent normal dispersions of the lattice vectors being verified experimentally using the function expansion by generalized spherical harmonics.

2. Fourier image of the theoretical distribution of orientations probability. The components of the distribution of orientations of the crystals are formed by the convolution of the functions on the group of continuous rotations. In Fourier space the convolution corresponds to the multiplication of matrices composed of spherical harmonics of functions to be convolved [100]:

$$W_{lmn}=\frac{1}{2l+1}\sum_{k=-l}^{l} Q_{lmk}R_{lkn}, \tag{1.4}$$

$$\begin{cases} Q_{lmn}=(2l+1)\int_{\mathbf{G}} Q(\mathbf{g})T^l_{mn}(\mathbf{g})d\mathbf{g},\\ R_{lmn}=(2l+1)\int_{\mathbf{G}} R(\mathbf{g})T^l_{mn}(\mathbf{g})d\mathbf{g}.\end{cases}$$

Calculation of spherical harmonics of the function $R(\mathbf{g},\mathbf{E})$, as shown by Eq. (1.2) reduces to averaging $T^l_{mn}(\mathbf{g})$ over all $\mathbf{E}$ orientations of the expected type $\mathbf{e}=\mathbf{g}(\psi_0,\vartheta_0,\varphi_0)$, appearing with the symmetrical rotations of crystal:

$$R_{lmn}=(2l+1)\left\langle T^l_{mn}(\mathbf{e})\right\rangle.$$

Fourier image of the function $Q(\mathbf{g},\mathbf{K})$ is sought in the form of a product of two integrals:

$$\eta_{lmn}=\frac{1}{2}\int_0^{\pi} e^{K_1\cos\vartheta}P_l^{mn}(\cos\vartheta)\sin\vartheta d\vartheta,$$

$$\tau_{mn}=\frac{1}{4\pi^2}\int_0^{2\pi}\int_0^{2\pi} e^{K_2\cos(\varphi+\psi)}e^{-i(m\psi+n\varphi)}d\varphi d\psi.$$

Generalized Legendre function $P_l^{mn}(t)$, where $t = \cos\vartheta$ with the existing limitation $|m| \le l$ is driven to a finite hypergeometric series provided that $m \ge n$ [24]:

$$P_l^{mn}(t) = \frac{1}{(m-n)!}\sqrt{\frac{(l+m)!(l-n)!}{(l-m)!(l+n)!}}\left(\frac{1-t}{2}\right)^{\frac{m-n}{2}}\left(\frac{1+t}{2}\right)^{\frac{m+n}{2}}\sum_{k=0}^{l-m}\lambda_k\left(\frac{1-t}{2}\right)^k,$$

$$\lambda_k = \frac{1}{k!}\frac{(-l+m)_k(l+m+1)_k}{(m-n+1)_k},$$

$$(a)_0 = 1,\ (a)_k = a(a+1)(a+2)\cdots(a+k-1),\ (1)_k = k!.$$

All emerging values of $P_l^{mn}(t)$ are calculated using the formula [100]:

$$P_l^{mn}(t) = (-1)^{m-n}P_l^{nm}(t),\ P_l^{mn}(-t) = (-1)^{l-m}P_l^{m\bar{n}}(t),$$

$$P_l^{mn}(1) = \delta_{mn},\ P_l^{mn}(-1) = (-1)^l\delta_{m\bar{n}}.$$

In the first integral we can substitute the expression of $P_l^{mn}(t)$, where $(1 \pm t)^k$ is resolved into components by Newton's binomial. Then η_{lmn} decomposes into a sum of integrals

$$Z_k(\mathrm{K}_1) = \int_{-1}^{1} t^k e^{\mathrm{K}_1 t}\,dt.$$

Calculate τ_{mn} it is easy by expanding the function under the integral on the modified Bessel functions [24]:

$$e^{\mathrm{K}_2\cos(\varphi+\psi)} = \sum_{k=-\infty}^{\infty} I_k(\mathrm{K}_2)\cos k(\varphi+\psi),$$

$$\tau_{mn} = \sum_{k=-\infty}^{\infty} I_k(\mathrm{K}_2)\delta_{mk}\delta_{nk} = I_n(\mathrm{K}_2)\delta_{mn}.$$

By form of τ_{mn} immediately it is clear that in the Eq. (1.4) after summation will remain only spherical harmonics Q_{lmm}. Their final formula takes the form

$$Q_{lmm} = \frac{2l+1}{2}\left[\left(\frac{\operatorname{sh} K_1}{K_1}\right)\left(\frac{I_0(K_2)}{I_m(K_2)}\right)\right]^{-1} \times$$

$$\times \sum_{i=0}^{l-m} \frac{\lambda_i}{2^i} \sum_{j=0}^{m} (-1)^j \binom{m}{j} \frac{1}{2^j} \sum_{k=0}^{i+j} (-1)^k \binom{i+j}{k} Z_k(K_1),$$

$$Z_k(\alpha) = \frac{e^{\alpha}}{\alpha}\left[1 + \sum_{r=1}^{k} (-1)^r \binom{k}{r} \frac{r!}{\alpha^r}\right] -$$

$$-(-1)^k \frac{e^{-\alpha}}{\alpha}\left[1 + \sum_{r=1}^{k} \binom{k}{r} \frac{r!}{\alpha^r}\right].$$

Theoretical harmonics of mixed distribution of crystal orientations, of the form (1.1), acquire the exact mathematical expression:

$$\hat{W}_{lmn}(\mathbf{B}) = \sum_{\nu=1}^{s} \mu_\nu \, Q_{lmm}(\mathbf{K}_\nu) \times \tag{1.5}$$

$$\times \left\langle \cos(m\psi_0 + n\varphi_0) \left\{ \frac{1}{2} \left[P_l^{mn}(\cos\vartheta_0) + (-1)^{l-m} P_l^{m\bar{n}}(\cos\vartheta_0) \right] \right\} \right\rangle_{\mathbf{E}_\nu}.$$

Angle brackets denote averaging over the cubic symmetry group. As a result, averaging spherical harmonics identically satisfy the conditions of invariance of the texture function as to symmetric rotations of crystal.

As it should be, the imaginary component of the Fourier coefficients $\hat{W}_{lmn} = \frac{1}{2}\left(\hat{U}_{lmn} + i\hat{V}_{lmn}\right)$ equal to zero, and $\hat{U}_{000} = 1$.

3. Generalized dispersion parameters of the orientations in the rhombic texture. A given on continuous rotation group the function $f(\psi, \vartheta, \varphi)$ has a one-valued decomposition into the function $\gamma_{mn}(0, \vartheta, 0)$ with fixed *m* and *n* presented in mutually orthogonal subspaces the space of $T_{mn}^{l}(\mathbf{g})$ [100]:

$$f(\psi, \vartheta, \varphi) = \sum_{m=-\infty}^{\infty} \sum_{n=-\infty}^{\infty} e^{-im\psi} \gamma_{mn}(0, \vartheta, 0) e^{-in\varphi},$$

$$\gamma_{mn}(0, \vartheta, 0) = \sum_{l=\max(|m|,|n|)}^{\infty} W_{lmn} P_l^{mn}(\cos\vartheta).$$

According to Parseval's equality,

$$\overline{\left|f(\psi,\vartheta,\varphi)\right|^2} = \sum_{m=-\infty}^{\infty}\sum_{n=-\infty}^{\infty}\overline{\left|\gamma_{mn}(0,\vartheta,0)\right|^2},$$

$$\overline{\left|\gamma_{mn}(0,\vartheta,0)\right|^2} = \sum_{l=\max(|m|,|n|)}^{\infty}\frac{1}{2l+1}\left|W_{lmn}\right|^2.$$

Matrix $\mathbf{\Gamma}$ composed of elements of $\overline{\left|\gamma_{mn}(0,\vartheta,0)\right|^2}$ associated with the vector $\mathbf{W}$ of spherical harmonics, gives a generalized description of the heterogeneity of orientations distribution [65].

Norm of the matrix $\mathbf{\Gamma}$ is divided into three parts:

$$\left\|\mathbf{\Gamma}\right\| = \left\|\mathbf{\Gamma}_0\right\| + \left\|\mathbf{\Gamma}_1\right\| + \left\|\mathbf{\Gamma}_2\right\| = \overline{\left|f(\psi,\vartheta,\varphi)\right|^2},$$

$$\begin{cases} \left\|\mathbf{\Gamma}_0\right\| = \left|W_{000}\right|^2 = \left|\overline{f}\right|^2, \\ \left\|\mathbf{\Gamma}_1\right\| = \overline{\left|\gamma_{00}(0,\vartheta,0)\right|^2} - \left|W_{000}\right|^2 = \overline{\left|\overline{f(\vartheta)} - \overline{f}\right|^2}, \\ \left\|\mathbf{\Gamma}_2\right\| = \sum\limits_{m=-\infty}^{\infty}\sum\limits_{n=-\infty}^{\infty}\overline{\left|\gamma_{mn}(0,\vartheta,0)\right|^2} - \overline{\left|\gamma_{00}(0,\vartheta,0)\right|^2} = \overline{\left|f(\psi,\vartheta,\varphi) - \overline{f(\vartheta)}\right|^2}, \end{cases}$$

here, $\overline{f}$ is the average of $f(\psi,\vartheta,\varphi)$ over all variables while $\overline{f(\vartheta)}$ is the average by ψ, φ at a given value of ϑ.

The mean square of fluctuations of the density distribution of orientations

$$\Pi = \left\|\Gamma_1\right\| + \left\|\Gamma_2\right\| = \overline{\left|f(\psi,\vartheta,\varphi) - \overline{f}\right|^2}$$

is a measure of closeness to the order: $\Pi = 0$ at random orientations of the crystals and $\Pi \to \infty$ in an ideal orientation order (δ-shaped fluctuation).

Parameters, which changing in the finite interval [0,1]

$$s_{hkl} = \frac{\left\|\Gamma_0\right\|}{\left\|\Gamma_0\right\| + \left\|\Gamma_1\right\|}, \quad s_{uvw} = \frac{\left\|\Gamma_0\right\| + \left\|\Gamma_1\right\|}{\left\|\Gamma_0\right\| + \left\|\Gamma_1\right\| + \left\|\Gamma_2\right\|}$$

characterize dispersion of the directions of the crystallographic vectors in plane deformed sample.

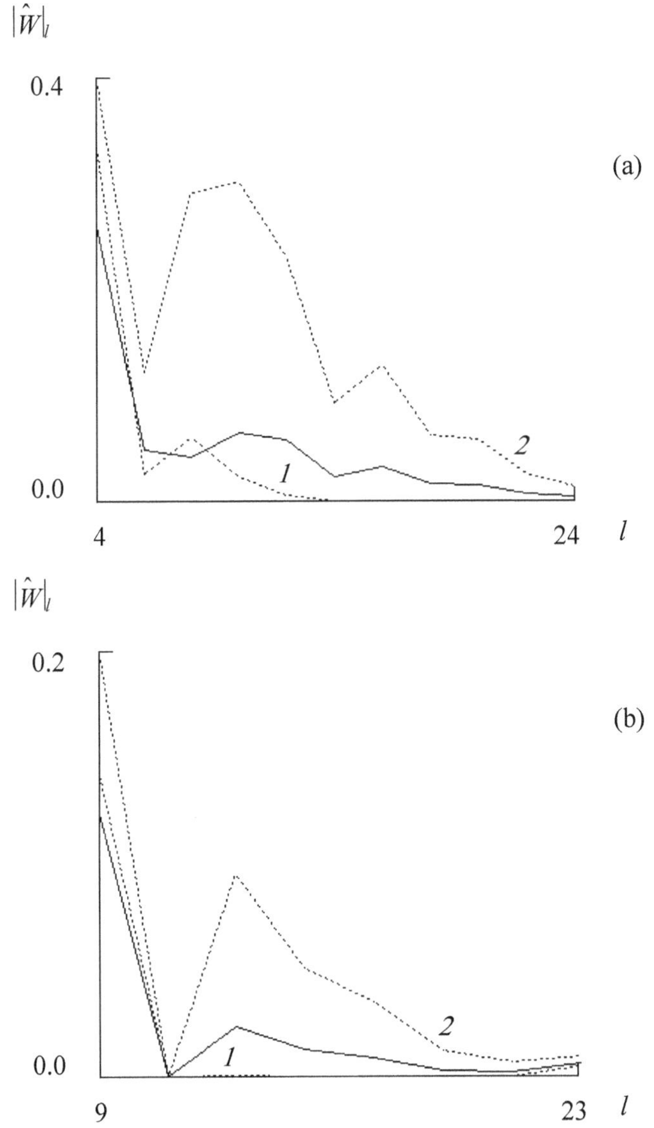

Fig. 1.1. Character of decreasing of the harmonics with an even (a) and odd (b) degree, depending on parameters of the texture function. The dashed lines for the components: (*1*) $(110)[001]$ ($K_1 = K_2 = 10$); (*2*) $(110)[1\bar{1}2]$ ($K_1 = K_2 = 50$). The solid line for the mixed distribution $(\mu_2/\mu_1 = 1/3)$

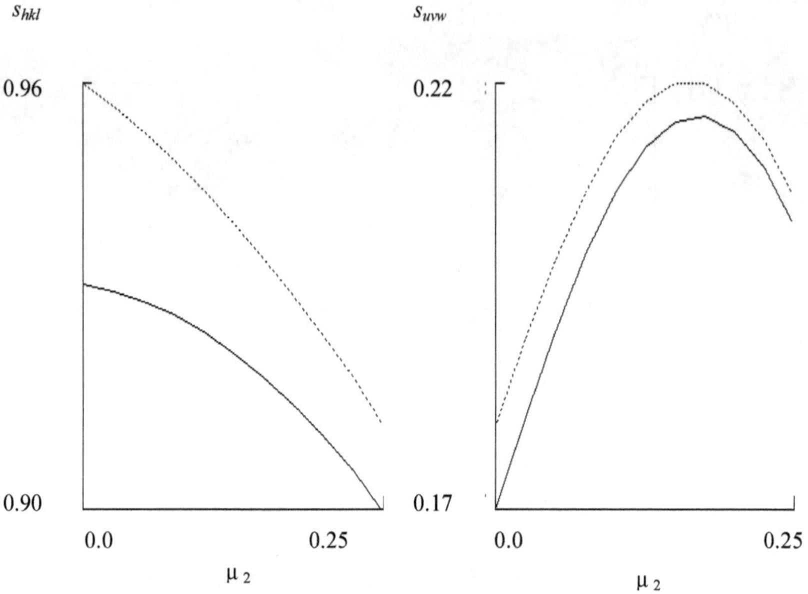

Fig. 1.2. Changing of of the generalized dispersion parameters of the orientations with an increase in the weight of sharp component of the texture function (parameters are the same as in Fig. 1.1). Solid lines by all harmonics of degree $l \leq 24$ and dashed lines only by harmonics of even degree

Disordering of the crystallographic planes (*hkl*) with respect to plane of deformation weakens the fluctuations by an angle ϑ to the Z axis, which coincides with the normal to plane: $\|\mathbf{\Gamma}_1\| \to 0$, $s_{hkl} \to 1$.

Disordering of orientations of the crystallographic directions [*uvw*], lying in the plane of the deformation weakens the fluctuations by an angle of rotation $(\varphi + \psi)$ around the Z axis: $\|\mathbf{\Gamma}_2\| \to 0$, $s_{uvw} \to 1$.

Generalized dispersion parameters of the orientations s_{hkl} and s_{uvw} depend on both coordinates and sharpness of the maxima of distribution density. Parameter $s_{hkl} > s_{uvw}$ when $K_1 = K_2$, even under ordering of crystal planes of type (001).

Taking as an example crystal orientations for texture of type silver, it can be observed decreasing of the averaged by modulus Fourier coefficients $\left|\hat{W}\right|_l$ with the increasing their degree l (Fig. 1.1), and an effect of mixing introduced distributions on the dispersion of orientations (Fig. 1.2).

Statistical model of the distribution of orientations reveals the crystal structure of metals being plane deformed. Problem is to determine parameters of the texture function by data of harmonic analysis.

CHAPTER 2
OPTIMAL DIFFRACTION EXPERIMENT FOR HARMONIC ANALYSIS OF THE TEXTURE FUNCTION

Uncertainty in the estimations of harmonics of the texture function reaches a minimum when being measured the most informative points in reciprocal space of a polycrystal, and the model of regression experiment contains only the significant harmonics subject to existing dispersion of the orientations of crystals [64, 83].

§ 2.1. Imaging of Distribution of Crystals Orientations in the Diffraction Intensity Fluctuations

Spherical harmonics of the texture function become the prototype of the Fourier representation of the observed intensity distribution of scattering by a polycrystal, from where regression estimates of desired harmonics to be excerpted.

1. The scattering property of a polycrystal with the existing distribution of orientations. The reciprocal space of a polycrystalline sample consists of concentric spheres formed by the rotation of the reciprocal crystal lattice around the lattice site taken as the zero. Symmetry of the reciprocal lattice coincides with the original lattice symmetry [98, 16].

From many spheres of the reciprocal space of crystals with cubic symmetry, there are three basic spheres described by the radius vectors of the crystal unit cell sites, that is [100], [110], [111].

Let $\mathbf{r} = (r, \psi, \vartheta)$ is a point on sphere of the reciprocal space of radius $r = \sqrt{h^2 + k^2 + l^2}$, to which is guided the diffraction vector $\mathbf{q}_{HKL}$. Fluctuation of the diffraction intensity $J(\mathbf{r})$ at point $\mathbf{r}$ is related to the density of orientations of crystallographic vectors $\langle hkl \rangle$ in the direction of vector $\mathbf{r}$, which depends on the texture function $f(\mathbf{g})$.

When the basis of a cubic crystal lattice coincides with the axes (X, Y, Z) of the external system of coordinates to make the vector $[hkl]$ agree with the Z axis, it is need lattice rotation $\mathbf{g}_{hkl}(\psi_{hkl}, \vartheta_{hkl}, \varphi_{hkl})$. There ϑ_{hkl} and φ_{hkl} are spherical coordinates of the vector $[hkl]$ in the crystallographic basis, and ψ_{hkl} is any angle of rotation around the $[hkl]$. So, for crystal with orientation $\mathbf{g}$ it is required the rotation of $\mathbf{g}_{hkl}\mathbf{g}^{-1}$.

Therefore, crystals in which the normal to the plane (hkl) coincides with the Z axis have the orientation $(\mathbf{g}_{hkl}\mathbf{g}^{-1})^{-1} = \mathbf{g}\,\mathbf{g}_{hkl}^{-1}$. The probability density of such orientations is

$$f\left(\mathbf{g}\,\mathbf{g}_{hkl}^{-1}\right)=f\left(\psi-\psi_{hkl},\vartheta-\vartheta_{hkl},\varphi-\varphi_{hkl}\right),$$

and with all possible rotations around the normal

$$\mathrm{P}_{hkl}(\mathbf{g})=\frac{1}{2\pi}\int_{0}^{2\pi}f\left(\psi-\psi_{hkl},\vartheta-\vartheta_{hkl},\varphi-\varphi_{hkl}\right)d\psi_{hkl}. \qquad (2.1)$$

The Fourier representation of the probability density distribution

$$f\left(\mathbf{g}\,\mathbf{g}_{hkl}^{-1}\right)=\sum_{l=0}^{\infty}\sum_{m=-l}^{l}\sum_{n=-l}^{l}W_{lmn}T_{mn}^{l}\left(\mathbf{g}\,\mathbf{g}_{hkl}^{-1}\right)$$

using properties of generalized spherical functions [100, 16]:

$$T_{mn}^{l}\left(\mathbf{g}\,\mathbf{g}_{hkl}^{-1}\right)=\sum_{\mu=-l}^{l}T_{m\mu}^{l}(\mathbf{g})T_{\mu n}^{l}\left(\mathbf{g}_{hkl}^{-1}\right),\; T_{\mu n}^{l}\left(\mathbf{g}_{hkl}^{-1}\right)=T_{n\mu}^{l}\left(\mathbf{g}_{hkl}\right)^{*}$$

can be reduced to the following form:

$$f\left(\mathbf{g}\,\mathbf{g}_{hkl}^{-1}\right)=\sum_{l=0}^{\infty}\sum_{m=-l}^{l}\sum_{n=-l}^{l}W_{lmn}\sum_{\mu=-l}^{l}\mathrm{A}_{m\mu}(\mathbf{g})\mathrm{B}_{\mu n}\left(\mathbf{g}_{hkl}\right), \qquad (2.2)$$

$$\begin{cases}\mathrm{A}_{m\mu}(\mathbf{g})=e^{-im\psi}\left\{\dfrac{1}{2}\left[P_{l}^{m\mu}(\cos\vartheta)+P_{l}^{m\mu}(-\cos\vartheta)\right]\right\}e^{-i\mu\varphi},\\ \mathrm{B}_{\mu n}\left(\mathbf{g}_{hkl}\right)=e^{i\mu\psi_{hkl}}P_{l}^{\mu n}\left(\cos\vartheta_{hkl}\right)e^{in\varphi_{hkl}}.\end{cases}$$

There is already provided invariance of $f\left(\mathbf{g}\,\mathbf{g}_{hkl}^{-1}\right)$ to the reflection of the normal to the sample plane along which the Z axis is directed.

In Equation (2.1) – (2.2) integrating with respect to ψ_{hkl} produces δ-function with μ. And summation over μ give rise to the following probability density for the orientation of the vector [*hkl*] in the Z axis:

$$\mathrm{P}_{hkl}(\mathbf{g})=\sum_{l=0}^{\infty}\sum_{m=-l}^{l}\sum_{n=-l}^{l}W_{lmn}\mathrm{A}_{m0}(\mathbf{g})\mathrm{B}_{0n}\left(\mathbf{g}_{hkl}\right),$$

$$\begin{cases}\mathrm{A}_{m0}(\mathbf{g})=e^{-im\psi}\left\{\dfrac{1}{2}\left[1+(-1)^{l-m}\right]\right\}P_{l}^{m0}(\cos\vartheta),\\ \mathrm{B}_{0n}\left(\mathbf{g}_{hkl}\right)=P_{l}^{0n}\left(\cos\vartheta_{hkl}\right)e^{in\varphi_{hkl}}.\end{cases}$$

If crystallographic vector $[hkl]$ is set as basis vector of the lattice, then function $\mathrm{P}_{hkl}(\mathbf{g})$, where $\mathbf{g} = \mathbf{g}(\psi, \vartheta, 0)$ shows the distribution density of random orientations of vectors $[hkl]$ by spherical coordinates ϑ and ψ in coordinates system of a sample designated (X, Y, Z).

The scattering property of a polycrystal in the direction of the vector $\mathbf{r} = (r_{hkl}, \psi, \vartheta)$ is the average $\mathrm{P}_{hkl}(\psi, \vartheta)$ for all vectors of type $\langle hkl \rangle$, appearing in the symmetry transformations of the crystal lattice, leaving unchanged diffraction pattern. Expression of $\langle \mathrm{P}_{hkl}(\psi, \vartheta) \rangle_{r_{hkl}}$ (the angle brackets denote averaging over the rotation sphere of radius r_{hkl}) becomes the equation for fluctuations of the scattering intensity on a polycrystal:

$$\left.\begin{aligned} &\hat{J}(\mathbf{r}) = \sum_{l=0}^{\infty} \sum_{m=-l}^{l} \sum_{n=-l}^{l} W_{lmn} X_{lmn}(\mathbf{r}), \\ &X_{lmn}(\mathbf{r}) = e^{-im\psi} P_l^{m0}(\cos\vartheta) \left\langle P_l^{0n}(\cos\vartheta_{hkl}) e^{in\varphi_{hkl}} \right\rangle_{r_{hkl}} \\ &(l = 0, 4, 6, \ldots;\ |m| = 0, 2, 4, \ldots, l;\ |n| = 0, 4, 8, \ldots, l). \end{aligned}\right\} \quad (2.3)$$

Harmonics of odd degree l have no effect on the fluctuations of the intensity in reciprocal space a polycrystal and consequently, in principle, are immeasurable.

Equation (2.3) contains the formula of surface spherical harmonics of a pole densities $J(r, \vartheta, \psi)$, which was received by R. Roe [56]:

$$\hat{J}(\mathbf{r}) = \sqrt{\frac{4\pi}{2l+1}} \sum_{l=0}^{\infty} \sum_{m=-l}^{l} S_{lm}(r) Y_{lm}(\vartheta, \psi),$$

$$S_{lm}(r) = \sqrt{\frac{4\pi}{2l+1}} \sum_{n=-l}^{l} W_{lmn} \left\langle Y_{ln}(\vartheta_{hkl}, \varphi_{hkl})^* \right\rangle_r .$$

Relationships between different spherical functions are given in § 1.1.

A surface spherical harmonic $S_{lm}(r)$ is a linear function of generalized spherical harmonics W_{lmn} of all orders n available for their degree l. To recover W_{lmn} with all allowable n at the same (l, m), it is necessary to have number of harmonics $S_{lm}(r)$ for the different r_{hkl} the higher, the higher l.

For crystals with cubic lattice there is one (W_{lm0} ($l < 12$)) or two (W_{lm0} and W_{lm4} ($12 \le l < 24$)) independent coefficient of the n-th order by number of invariants with respect to symmetrical lattice rotations (Table 1.1). And more there is an independent coefficient W_{000} being invariant with respect to all rotation of the three-dimensional space.

It follows that the necessary and sufficient number of spheres of different radius r, which must be present in the observation region of the reciprocal space of cubic crystals, is equal to two or three respectively specified limit of harmonics degree $l < 24$.

If a single sphere of observations, even when $l \leq 10$, coefficients W_{lmn} of nonzero degree become strongly correlated with the outweighing coefficient W_{000} and their estimations acquire large systematic errors [83].

2. The regression model of fluctuations of the scattering intensity at real texture of a polycrystal. In fact, the highest degree of significant coefficients of the texture function expansion is substantially limited.

The condition of convergence of the Fourier series for the texture function $f(\mathbf{g})$, which is given in § 1.1 can be written as

$$\sum_{l=0}^{\infty}(2l+1)^{-1}\sum_{m=-l}^{l}\sum_{n=-l}^{l}\left|W_{lmn}\right|^{2} < +\infty.$$

It is known that a series $\sum_{l=0}^{\infty} l^{-q}$ converges for $q > 1$ and diverges for $q \leq 1$ [37]. Consequently, if a series representing the $f(\mathbf{g})$ generally converges, then $(2l+1)^{2}\left|W\right|_{l}^{2}$, where $|W|_l$ denotes the average of all $|W_{lmn}|$ with fixed l, decreases not slower than $l^{-\omega}$ ($\omega > 0$).

Let us assume that $|W|_l$ measurement errors do not increase with l. The highest degree of reliable harmonics $l_{\max}$ will be determined by the inequality $2\left|W\right|_{l} \geq \sigma_W$, where σ_W denotes the error of the normalized coefficients $|W|_l$. Assuming that $2\left|W\right|_{l} \sim l^{-\frac{1}{2}(2+\omega)}\left(2l \gg 1\right)$, there is

$$l_{\max} \leq e^{-\{2\ln\sigma_W/(2+\omega)\}} \quad \left(\sigma_W \leq 1\right).$$

The dependence of the highest degree of significant coefficients W_{lmn} on the parameter ω, which is a measure of the imperfection of the crystallographic texture, is shown in Fig. 2.1.

At small measurement errors σ_W, a great role plays minor deviations from the ideal texture. With increasing σ_W the highest degree of reliable harmonics $l_{\max}$ immediately drops sharply, further continuing to fall under weakening of texture.

When actually there is a mixed distribution of orientations, error σ_W increases due to fluctuations of a random distribution of orientations with aggregate of scattering crystals [74].

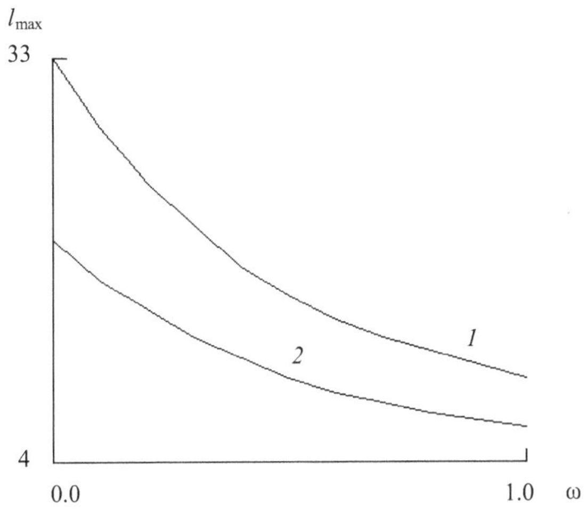

Fig. 2.1. Degree of reliable harmonic of the texture function at situation of the orientations dispersion, and measurement errors: *(1)* $\sigma_W = 0.03$; *(2)* $\sigma_W = 0.05$

In the infinite-dimensional space of spherical functions for observed the diffraction intensities we need to build the best regression model $J(\mathbf{r},\mathbf{W})$, where the vector of parameters $\mathbf{W} = \{W_{lmn}\}$ contains the greatest number of reliable harmonics of the texture function.

§ 2.2. Measurement of Harmonics of the Texture Function in Reciprocal Space of a Polycrystal

Number received in the experiment information about the parameters of a linear regression model in advance can be maximized by constructing a good covariance matrix of estimates.

1. The best linear regression estimate for the Fourier coefficients of the density distribution of orientations. An unbiased estimate for infinite-dimensional true vector of harmonics $\hat{\mathbf{W}}$ does not exist. The best of its biased estimates by regression model $J(\mathbf{r},\mathbf{W})$, in the sense of the lowest possible standard deviations from $\hat{\mathbf{W}}$, can only be vector $\mathbf{W}$ with minimal norm $\|\mathbf{W}\|^2 = \mathbf{W}^t\mathbf{W}$ (t-superscript denotes transpose). Limiting a bias in estimates of $\mathbf{W}$ is equivalent to their stabilization [49, 95].

Suppose that the vector of measurements $\mathbf{J}$ has the covariance matrix $\mathbf{V} = \sigma^2 \mathbf{I}$, where σ^2 is the variance of the measurements, $\mathbf{I}$ is unit $(M{\times}M)$ matrix, M is dimension of the vector $\mathbf{J}$. The dimension of being estimated vector $\mathbf{W}$, denoted N depends on the highest degree Fourier coefficients l_{max} in the regression equation $J(\mathbf{r},\mathbf{W})$.

Optimal by accuracy and stability the estimate of $\mathbf{W}^* = \{W_\mathbf{l}^*\}$ minimizes the functional

$$Q(\mathbf{W}) = [\mathbf{RW} - \mathbf{J}]^t [\mathbf{RW} - \mathbf{J}] + \omega \mathbf{W}^t \mathbf{W},$$

$$\begin{cases} \mathbf{R} = \{R_{\mu\nu}^{lmn}\},\ R_{\mu\nu}^{lmn} = X_{lmn}(r_\mu, (\psi, \vartheta)_\nu), \\ \mathbf{J} = \{J_{\mu\nu}\},\ J_{\mu\nu} = J(r_\mu, (\psi, \vartheta)_\nu) \\ \quad (\mu = 1,\ \ldots,\ q\ (\nu = 1,\ \ldots,\ M_\mu)). \end{cases}$$

Matrix **R** is constructed from basis functions of the Fourier representation (2.3), calculated at points of measurements; M_μ is the number of points on μ-th a sphere, q is number of spheres; $\sum_{\mu=1}^{q} M_\mu = M\ (M > N,\ q > 1)$.

Minimizing the norm of vector **W**, regulated by a parameter $0 < \omega \le 1$, leads to the choice of the most smooth regression function $J(\mathbf{r},\mathbf{W})$ of all consistent with the data of measurements **J** within their variance σ^2. The estimate of vector W becomes resistant to a random fluctuation of data sample [95].

In "regularized" regression the estimate is calculated by the formula

$$W_\mathbf{l}^* = C_{\mathbf{ll}'}^{-1} \sum_{\mu=1}^{q} \sum_{\nu=1}^{M_\mu} R_{\mu\nu}^{\mathbf{l}'} J_{\mu\nu},\quad C_{\mathbf{ll}'} = \sum_{\mu=1}^{q} \sum_{\nu=1}^{M_\mu} R_{\mu\nu}^{\mathbf{l}} R_{\mu\nu}^{\mathbf{l}'} + \omega\, \delta_{\mathbf{ll}'}.$$

Here, $C_{\mathbf{ll}'}^{-1}$ are the elements of the inverse matrix $\mathbf{C}^{-1}$; $\delta_{\mathbf{ll}'}$ is Kronecker symbol ($\mathbf{l} = (l, m, n)$).

If agreed between each other an error of the regression model, variance of measurements and regularization parameter: $\hat{\mathbf{W}}^t \hat{\mathbf{W}} \le \omega / \sigma^2$, then the estimate $\mathbf{W}^*$ has the lowest generalized variance, or the smallest determinant of the covariance matrix det $\mathbf{V_W}$, where $\mathbf{V_W} < \sigma^2 [\mathbf{R}^t \mathbf{R}]^{-1}$ [49].

The bias of estimate of vector **W** as a result of errors of the model $J(\mathbf{r},\mathbf{W})$ is proportional to the matrix $\mathbf{C}^{-1}$. Thus, estimate with the smallest generalized variance $\sim \det \mathbf{C}^{-1}$ has the smallest the bias. Therefore will be the lowest possible standard deviations of $W_\mathbf{l}^*$ from the true values $\hat{W}_\mathbf{l}$.

A measure of uncertainty in the estimation of $W_\mathbf{l}^*$ this is $\log \det \mathbf{V_W}$. The amount of information obtained in the experiment is equal to reducing uncertainty. Measurement points $(\mathbf{r}_1,\ \ldots,\ \mathbf{r}_M)$ must be selected so as to minimize $\log \det \mathbf{V_W}$. As shown in [2], the minimization of $\log \det \mathbf{V_W}$ or, equivalently, $\det \mathbf{V_W}$ is equivalent to maximizing $\det\left[\mathbf{V} + \mathbf{RV_W R}^t\right]$.

Diagonal elements of the matrix $\mathbf{RV_W R}^t$ when the covariance matrix of the measurement errors $\mathbf{V} = \sigma^2\mathbf{I}$ are of the form $\sigma^2 d(\mathbf{r})$, where

$$d(\mathbf{r}) = \sum_{l}\sum_{l'} X_l(\mathbf{r})\, C_{ll'}^{-1}\, X_{l'}(\mathbf{r}),$$

and they are an expression of the variance of estimates of the output variable $J(\mathbf{r},\mathbf{W})$ at the point $\mathbf{r}$.

Thus, when are measured those points $(\mathbf{r}_1, \ldots, \mathbf{r}_M)$, where the error of the predicted values of $J(\mathbf{r},\mathbf{W})$ is the highest, the most amount of useful information is acquired [2].

2. Uniformly the most informative points of measurement in the reciprocal space of a polycrystal. Experiment for estimation of parameters of inaccurate regression models can be optimized in the same way as if the exact model. Although minimality of the highest error in the estimates of itself regression function will not be fulfilled [49].

For usual linear regression we have $\mathbf{V_W} = \sigma^2\mathbf{C}^{-1}$. The criterion of experiment optimality is min det $\mathbf{V_W}$. It is therefore necessary to either minimize det $\mathbf{C}^{-1}$ or maximize det $\mathbf{C}$, where $\mathbf{C} = [\mathbf{R}^t\mathbf{R}]$ is the information matrix.

Matrix $\mathbf{D} = \mathbf{C}^{-1}$ is called a dispersion matrix of a design of experiment specified by $E = (\mathbf{r}_1, \ldots, \mathbf{r}_M; \xi_1, \ldots, \xi_M)$, where ξ_j is the weight of j-th measurement point, $\sum_{j=1}^{M} \xi_j = 1$. The design E^*, satisfying criterion of minimum det $\mathbf{D}$, there is the D-optimal design.

The problem of constructing the D-optimal design is solved by searching for the global extremum of the objective function is uniquely associated with the selected optimality criterion.

Always there is a D-optimal design E^* with a finite number of points

$$N \le M \le N(N+1)/2,$$

where N is the dimension of the vector of parameters of regression model (Kiefer, Fedorov). A necessary and sufficient condition for D-optimality of design E^* is the equality of

$$\max d(\mathbf{r}, E^*) = N$$

for all points of $\mathbf{r}$ in the designing area Ω (Kiefer and Wolfowitz) [49].

Best measurement points to estimate the harmonics of the texture function by the regression model of diffraction intensity fluctuations are found using the algorithm of accelerated search D-optimal designs [43].

With accelerated method to the original plan $E^{(i)}$ (i = 0, 1, 2, …) is added at once many such points from the available measurement region Ω, where $d(\mathbf{r}, E^{(i)})$ reaches a maximum. Weight ξ_j, which is assigned to point $\mathbf{r}_j$ (j = 1, 2, … , $M^{(i)}$) , depends on how $d(\mathbf{r}, E^{(i)})$ exceeds N. The points with low weight ξ_j gradually are superseded in an iterative process.

Sequence of designs $E^{(i)}$ (i = 0, 1, 2, …) has being proven convergence to D-optimal design E^*. In a practice, optimization is completed when

$$\frac{\det \mathbf{C}\left(E^{(i+1)}\right)-\det \mathbf{C}\left(E^{(i)}\right)}{\det \mathbf{C}\left(E^{(i+1)}\right)}<\varepsilon_1, \quad \frac{\max d\left(\mathbf{r}, E^{(i+1)}\right)-N}{N}<\varepsilon_2;$$

ε_1 and ε_2 are as acceptable errors.

Initial designs for the regression models $J(\mathbf{r},\mathbf{W})$ with the specified dimension of the vector $\mathbf{W}$ included a large number of points $\left(\mathbf{r}_1, \ldots, \mathbf{r}_M\right)$, where $M \gg N$, distributed randomly for the cubic crystals into their three basis spheres in the diffraction space. Designs [64] approaching to the D-optimal, contain the minimal number of points $M = N$ with approximately equal weight $\xi_j \approx M^{-1}$ (j = 1, … , M).

In Figure 2.2 are shown examples of optimal arrangement of observations in the reciprocal space of a polycrystal at different highest degree of harmonics of the texture function. Points $\left(\mathbf{r}_1, \ldots, \mathbf{r}_M\right)$ symmetrically duplicated on spheres for randomization of measurements.

By optimizing almost uniformly the most informative observation points are disposed exactly at the necessary and sufficient number of spheres for cubic crystals. Density of points increases rapidly with the approach to the border of observations region in the reciprocal space.

For approximation of pole figures {hkl} using the surface spherical harmonics $Y_{lm}(\vartheta,\psi)$ (q = 1) there is optimal the uniform arrangement of points over area of the sphere [78]. The result is as like obvious: for the harmonic analysis of the diffraction line is optimal uniform arrangement of points on the interval $(-\pi, \pi)$ [49] . Adequate representation of the function specified on a sphere requires a much smaller number of observation points because each point has already two coordinates.

Now is uncovered how traditional measuring of pole figures are unsuccessful for harmonic analysis even of themselves pole figures, and the more the texture function. On a regular grid of angles of the spherical coordinate occur gathering of points to a pole, furthermore with the increasing radius of a sphere in the reciprocal space the points average density decreases, instead of rising.

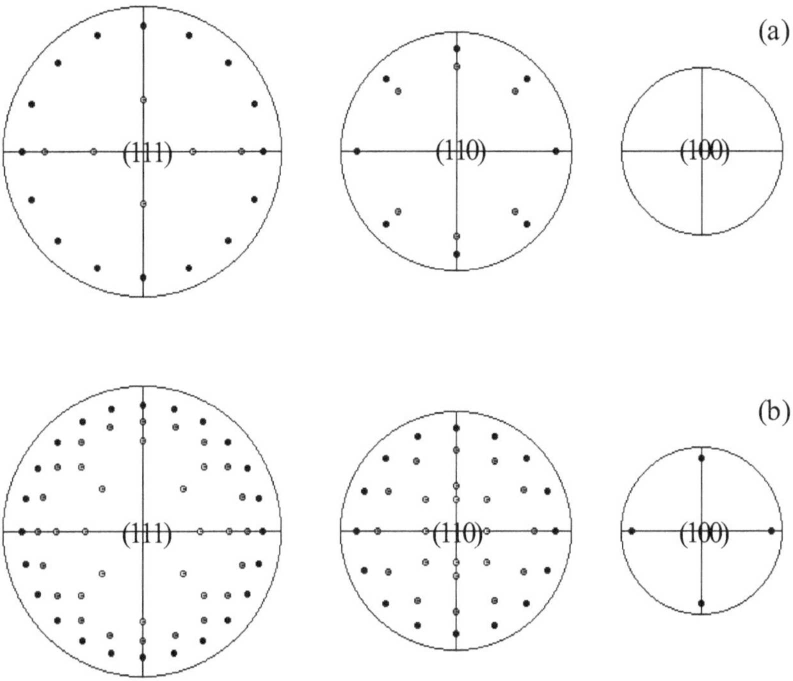

Fig. 2.2. Optimal arrangement of the measuring points for estimation of harmonics of the texture function: (a) $l_{max} = 8$, (b) $l_{max} = 12$ ($\vartheta \leq 60°$)

Very poorly organized measurements of pole figures have different and correlated errors. Into fitting coefficients an uncertainty which is generated by uncontrolled errors is so great that the true harmonics of the texture function actually do not known. Traditional experiment does not add information about texture function to an existing a priori.[1]

When optimal experiment the most effective measurements for estimation of harmonics of the texture function are performed with the highest reliability.

3. Measurement data with the best cooperative covariance matrix. Criterion of measurement quality it is the independent identical error distribution with the mathematical expectation equal zero (systematic errors are absent).

---

[1] Function approximation of pole figures by "Gaussian peaks" [89] further from physical reality, since does not already contain relating thereto a priori information.

With optimal use of observation area Ω the data of measuring points $(\mathbf{r}_1, \ldots, \mathbf{r}_N)$ are independent. The scattering volumes, because of the divergence of the beam of rays and the dispersion of wavelengths anywhere in the reciprocal space do not overlap. Changes in the size and shape of these scattering volumes at different points of reciprocal space, if they are significantly smaller than lattice sites are separated by background, for the observed of scatter intensity distribution are inconsequential [10].

The measured intensity of diffraction on a polycrystal by nature is a random variable with a Poisson composite distribution. Variance of normalized intensity mostly consists of relative fluctuations of random number of crystals within effective scattering volume of a sample [34] and of the proportion of crystals at reflecting position under random distribution of orientations [83]. Fluctuations of the number of crystals, illuminated by a beam of rays, do not so much depend on the coordinates of points $\mathbf{r}$ in reciprocal space as fluctuations of the orientations distribution which inherent to a multi-component texture [74].

Upon randomized procedure Poisson's variance of the measured quantity is part of a sample variance.

Randomization prevents from systematic distortion of data due to instrument drift or inaccuracy the geometry of diffraction with displacement of sample plane. The effect of these factors, as well as errors in determining the axes of the sample symmetry turned into a random.

The algorithm of the measuring control in optimal experiment [80]:

1. Choose a random point from set of the design points $(\mathbf{r}_1, \ldots, \mathbf{r}_N)$.

2. Set the measuring parameters of point $\mathbf{r}_j$: interval of angles of reflection $\Delta 2\theta(r_{hkl})$ to measure the whole intensity of above background; the angles of inclination and rotation of a sample relative to the reflected beam namely $\beta = \vartheta$ and $\alpha = \psi$.

3. Measure the intensity of reflection integrated when the counter moving in the interval $\Delta 2\theta(r_{hkl})$, and of background on the ends of the interval.

4. Duplicate the measuring of point $\mathbf{r}_j$ with symmetrical rotations of a sample: $(-1)^i \alpha + k\pi$, where (i, k) are take random integer values (0, 1).

5. Repeat steps *1 – 4* until all of N points in the design will be measured.

6. Replace a sample and repeat entire the measuring program.

Defects of the "slit device" that distort the specified coordinates of points $(\mathbf{r}_1, \ldots, \mathbf{r}_N)$ should be removed [23].

Optimization of time division when measuring the scattering intensity: An estimate of variance of the weighted average $\overline{\overline{J}}_\mu$ of all measured relative intensities on μ-th sphere when the total measuring time T_μ is $s_\mu^2 \approx \left(\overline{\overline{J}}_\mu T_\mu\right)^{-1}$. From the point of view of the uniformity Poisson's variances the best division of total measuring time $T_\Sigma = \sum T_\mu$ on different spheres μ will that in which $\left(\overline{\overline{J_\mu T_\mu}}\right) \approx \text{const}$.. Minimal required measuring time T_μ is determined by the allowable value of variance $s_\mu^2 \ll 10^{-4}$.

Getting information from weak reflections will require more time to measuring.

By the condition of minimizing the error of background subtraction, time of its measuring under the total time T_μ should make up, as shown in [105],

$$T_{b\mu} = \frac{\sqrt{\gamma_\mu}}{1+\sqrt{\gamma_\mu}} T_\mu,$$

where γ_μ is the relative level of the background at the μ-th sphere. The higher level of background, the more time required for its measuring.

The average time of one measuring on each of two samples

$$\overline{t_\mu} = \frac{1}{2}\frac{T_\mu}{pN_\mu}, \quad \overline{t_{b\mu}} = \frac{1}{2}\frac{\sqrt{\gamma_\mu}}{1+\sqrt{\gamma_\mu}}\overline{t_\mu}.$$

Here, N_μ is number of measuring points at the μ-th sphere; p is number of replicate measuring upon symmetric rotation of a sample.

Statistical estimation of primary experimental data: Data for harmonic analysis of the texture function this is measured normalized intensities from the level of background namely $J_{j,k}$ ($j = 1, \ldots, N$ ($k = 1, \ldots, p$)). Normalization is performed by an average intensity above background measured on the reference sample with the nearly randomly oriented crystals. For each μ-th sphere on the reference sample is measured large number of points with rapid rotation around the normal to the plane of a sample. The variance of the mean intensity is negligible compared to the variance at one point.

Estimates of sample mean values of $\overline{J_j} = \overline{J}(\mathbf{r}_j)$ and variances of s_j^2 are calculated with weight of w_k when the same number p of repeated measuring of each j-th point:

$$\overline{J_j} = \sum_{k=1}^{p} w_k J_{j,k}, \quad s_j^2 = \sum_{k=1}^{p} w_k \left[J_{j,k} - \overline{J_j}\right]^2, \quad w_k = J_{j,k}^{-1}\left[\sum_{k=1}^{p} J_{j,k}^{-1}\right]^{-1}.$$

Measuring in the conditions of randomization increases a variation in data. Repeated experiments make improving the accuracy of sample means.

When large expected values of $J(\mathbf{r}_j)$ the Poisson's their distribution comes near to the normal one. To test the homogeneity of variance of measurements $J(\mathbf{r}_j)$ by sample estimate of s_j^2 ($j = 1, \ldots, N$) can be used approximate criterion

$$\eta = N(p-1)\ \ln\sum_{j=1}^{N}\left(s_j^2/N\right) - (p-1)\sum_{j=1}^{N}\ \ln s_j^2,$$

which has an approximately χ^2-distribution with $(N - 1)$ degrees of freedom [31].

A check on study samples does not reject the assumption of statistical homogeneity of standard deviations in the region of observations. Therefore, the total variance of measurements can be estimated by pair differences of data for two samples [62]:

$$s^2 = \frac{1}{2N}\sum_{j=1}^{N}\left[\overline{J}_j^{(1)} - \overline{J}_j^{(2)}\right]^2.$$

Two series of independent measuring at each point of the design $(\mathbf{r}_1, \ldots, \mathbf{r}_N)$ give the vector of observations $\mathbf{J}$ of dimension $2N$ sufficient for simultaneous estimation of regression model $J(\mathbf{r},\mathbf{W})$ and checking its agreement with the data. The vector $\mathbf{J}$ has covariance matrix of errors distribution approaching to that in which will be a minimal volume of the dispersion ellipsoid of the estimates as $\sqrt{\det \mathbf{V}_\mathbf{W}}$ [2].

4. Robust model of the regression experiment for measurement of harmonics of the texture function. Degree harmonics of the texture function included in the regression model should be not less than required under the occurring sharpness of texture, but also not greater than available to measuring at existing experimental accuracy.

Economical of mathematical model by the number of estimated parameters is the condition of accuracy and stability of estimates [5]. The principle of economy is consistent with the requirement of a minimum norm of the vector of parameters for resistance to inaccuracies of model and data [95].

From the class of linear regression models $J(\mathbf{r},\mathbf{W})$ for which

$$\left\| J(\mathbf{r},\mathbf{W}) - \mathbf{J} \right\|^2 \le \boldsymbol{\delta}^t\boldsymbol{\delta} + \operatorname{Tr}\mathbf{V},$$

where $\boldsymbol{\delta}$ is the vector of the model errors in observation points, $\mathbf{V}$ is covariance matrix of errors of the measurements $\mathbf{J}$, it is necessary to choose model of the smallest order of N. It is in such model the vector of parameters will have minimal norm $\|\mathbf{W}\|^2$.

Determinant of the dispersion matrix $\det\mathbf{D}$ increases with its dimension $(N\times N)$. To limit the generalized variance of $\mathbf{W}$ estimate when increasing dimension of N, growth of $\det \mathbf{D}$ should be compensated by a decrease of the measurement variance σ^2. Otherwise, the increase in uncertainty in the quick-growing value of $\sigma^2 \det \mathbf{D}$ leads to the degeneration of the regression model, there does not remain reliable no single parameter.

Vector **e** of the bias in estimate of regression parameters of **W** with covariance matrix $\mathbf{V_W}$ satisfies equation

$$\mathbf{e}^t\,\mathbf{V_W}^{-1}\,\mathbf{e} = \left(\boldsymbol{\delta}^t\boldsymbol{\delta}\right)/\sigma^2.$$

For the regression model, which comparable in accuracy with the data, so that the decreasing of errors $\boldsymbol{\delta}$ through rise the number of Fourier coefficients is accompanied by a decrease of measurement errors σ, value $\left(\boldsymbol{\delta}^t\boldsymbol{\delta}\right)/\sigma^2$ is approximately constant. In these circumstances the covariance matrix of estimate $\mathbf{V_W}$ and the vector of bias **e** being minimized in concert.

Before the experiment, neither the sharpness of texture, which determines the rate of decreasing of the harmonics nor variance of measurements σ^2 is not known. Choosing economical model $J(\mathbf{r},\mathbf{W})$, where only parameters being estimated as needed for agreement with the data, naturally to start with the lower of highest-degree of harmonics l_{max} which defines the model order of N (Table 1.2).

Rising of l_{max} makes sense only until the accuracy of the approximation of observations **J**, estimated by the mean square residual deviations from $J(\mathbf{r},\mathbf{W})$ longer is no improving. If agreement with data when checking on the F-criterion with a given confidence probability P is not rejected, found estimates of **W** can be considered optimal.

At the beginning of the study to choose the best regression model it will be good practice to carry out the experiment with additional points to D-optimal design in the pole of spheres included into the observations area, to do the probability of a mistake in decision on the adequacy of the model as little as possible.

Sequential strategy in practice provides optimum choice of a regression model, parameters of which are the most accurately measured harmonics of the texture function.

§ 2.3. Practical Application of the Methods for Measuring the Orientation Distribution of Crystals

The actual quality of estimates of the Fourier coefficients of the orientation distribution density with the existing crystallographic texture is revealed by testing.

1. A limited class of regression models for practical texture analysis. To model of the regression experiment should be immediately excluded practically inaccessible under the achievable measuring accuracy, the highest terms of the Fourier series (2.3).

Experience shows that to describe the fluctuations of the diffraction intensity with accuracy not below than the one with which they can be measured, absolutely are sufficient Fourier coefficients W_{lmn} not more than tenth degree. When $l \leq 10$ there is independently only W_{lmn} of the order of $n = 0$. Moreover, on the invariance condition of output quantity with respect to symmetric rotations in the plane of a sample, the imaginary component of the Fourier coefficients of $W_{lmn} = \frac{1}{2}\left(U_{lmn} + iV_{lmn}\right)$ equal to zero.

Applicable in practice model of the regression experiment takes the form

$$\left.\begin{aligned} J(\mathbf{r},\mathbf{W}) &= U_{000} + \sum_{l=4}^{l_{\max}}\sum_{m=0}^{l} U_{lm0}X_{lm}(\mathbf{r}), \\ X_{lm}(\mathbf{r}) &= K_l H_l(r_{hkl})\cos m\psi\sqrt{\frac{(l-m)!}{(l+m)!}}P_l^m(\cos\vartheta), \\ \mathbf{r} &= (r_{hkl},\psi,\vartheta), \quad r_{hkl} = \sqrt{h^2+k^2+l^2}. \end{aligned}\right\} \qquad (2.4)$$

Coefficients U_{lm0} of order $m = 0$ are entered with the weight of $\frac{1}{2}$.

The function $H_l(r_{hkl})$ consists of Legendre polynomials $P_l(\cos\vartheta_{hkl})$:

$$H_l(r_{hkl}) = \frac{1}{3}\left[P_l\left(l/r_{hkl}\right) + P_l\left(k/r_{hkl}\right) + P_l\left(h/r_{hkl}\right)\right].$$

Coefficients K_l contain numeric constants which relate spherical harmonics with cubic symmetry of crystals:

$$K_l = 1 + 2\sum_{n=4}^{l}\zeta_{ln}^2 \quad (l \leq 10),$$

$$\langle Y_{ln}\rangle = \zeta_{ln}\langle Y_{l0}\rangle, \quad \langle T_{mn}^l\rangle = \zeta_{ln}\langle T_{m0}^l\rangle \quad (n = 4,\ 8,\ \ldots,\ l);$$

angle brackets, as before, denote averaging over the symmetry group.

2. Validating of the estimation accuracy of harmonics of the texture function using simulation experiments. By modeling of measurements in the reciprocal space of a polycrystal can be seen the deviation of estimates from true harmonics of the texture function and ascertain how affect the quality of estimates the option of model of the regression experiment, arrangement of the points in the observations area and measurement variance [83].

To imitative harmonic analysis of the texture function was accepted a theoretical probability distribution of orientations with the same components as in Fig. 1.1:

Component	*Type*	*Weight*	*Parameters*
$\nu = 1$	$\{110\}\langle 100\rangle$	$\mu_\nu = 0.85$	$K_{\nu 1} = K_{\nu 2} = 10$
$\nu = 2$	$\{110\}\langle 112\rangle$	$\mu_\nu = 0.15$	$K_{\nu 1} = K_{\nu 2} = 50$

Data of texture measurements at points of *D*-optimal design $(\mathbf{r}_1, \ldots, \mathbf{r}_N)$ are simulated by normally distributed a random variables with mathematical expectation of $\hat{J}(\mathbf{r})$ and covariance matrix $\mathbf{V} = \sigma^2 \mathbf{I}$, where σ^2 is variance of measurement at a point, $\mathbf{I}$ is unit $(N \times N)$ matrix, N is the number of measuring points.

Theoretical the pole densities of $\hat{J}(\mathbf{r})$ for generating of data are calculated using harmonics of the texture function $\hat{U}_{lmn}$ (1.5). Approximation is sufficiently accurate: the calculated values of $\hat{J}(\mathbf{r})$ are stable up to the sixth decimal place at all points of observations $(\mathbf{r}_1, \ldots, \mathbf{r}_N)$ under sequential increase the highest-degree of harmonics $24 \le l_{\max} \le 26$.

Each of simulation experiment produces a vector of observations $\mathbf{J}$ of dimension $2N$. From the data obtained are calculated the best estimates of parameter vector $\mathbf{U}$ of the regression model (2.4) as described in § 2.2. Experiments are repeated many times to retrieve sampling of the measured vectors $\mathbf{U}$ of the given volume M.

The true mean square error of estimates of the Fourier coefficients U_{lm0}:

$$\hat{\sigma}_{lm} = \sqrt{\frac{1}{M}\sum_{j=1}^{M}\left|U_{lm0}^{(j)} - \hat{U}_{lm0}\right|^2}$$

it is summation of their variation relative to the average of $\overline{U}_{lm0}$ and of deviation of the average $\overline{U}_{lm0}$ from the true value $\hat{U}_{lm0}$.

Table 2.1 shows the estimates of errors over random samples of size $M = 60$ when variance of the simulation measurements of $\sigma^2 = 0.01$.

Standard deviations of the measured normalized harmonics U_{lm0} from the true values $\hat{U}_{lm0}$ are compared on two models of the regression experiment with the highest degree of harmonics $l_{\max} = 8$ or 10.

Table 2.1.

Estimates of harmonics of the texture function by data of simulation experiments

Indexes $l\ m\ n$	*True value* $\hat{U}_{lm0}$	*The sample mean of estimates* $\overline{U}_{lm0}$		*Standard deviation from true value* $\hat{U}_{lm0}$	
		$l_{max} = 8$	$l_{max} = 10$	$l_{max} = 8$	$l_{max} = 10$
4 0 0	–0.6919	–0.8674	–0.8004	0.2034	0.1302
4 2 0	–2.8390	–2.7454	–2.8627	0.1247	0.1899
4 4 0	1.2260	1.0958	1.0105	0.1465	0.2265
6 0 0	–0.2486	–0.1989	–0.2065	0.0639	0.0640
6 2 0	0.1709	0.1498	0.1512	0.0249	0.0264
6 4 0	0.2023	0.1809	0.1666	0.0256	0.0395
6 6 0	0.1653	0.1345	0.1713	0.0444	0.0314
8 0 0	0.2788	–0.1219	0.4373	0.4232	0.3849
8 2 0	0.2631	–0.0779	0.0751	0.3533	0.2448
8 4 0	–0.1399	–0.0043	0.0610	0.1466	0.2122
8 6 0	0.1787	–0.2494	–0.2708	0.4336	0.4580
8 8 0	0.2056	–0.0593	0.0732	0.3054	0.4905
10 0 0	0.0056		0.0745		0.0988
10 2 0	0.0369		0.0615		0.0495
10 4 0	0.1128		–0.1219		0.2546
10 6 0	0.1722		0.0814		0.1071
10 8 0	0.0032		0.0007		0.0391
10 10 0	0.2067		0.1463		0.0860

With increasing the dimension of the vector of regression parameters **U** mean square deviation from the measurement data remains at the level of their variance σ^2, while the uncertainty and instability of the parameter estimates is rising rapidly. The degree of instability is seen in variations of maximum relative error $\hat{\sigma}_{lm} / \hat{U}_{lm0}$ on random samples of data, which are registered in Table 2.2.

Table 2.2.

Fluctuations of the maximum errors in the estimates of harmonics over repeated samples of simulation measurements data

The sample number	*The relative standard deviation*		*The relative bias of sample mean*	
	$l_{max} = 8$	l_{max}=10	$l_{max} = 8$	$l_{max} = 10$
1	2.29	15.3	–2.40	10,1
2	2.29	16.7	–2.32	10,6
3	2.36	17.7	–2.33	12,3
4	2.29	16.4	–2.33	10,4
5	2.34	12.4	–2.35	6,5

Error of the model with the $l_{max} = 8$ is more, and it caused the bias lowering estimates of higher harmonics U_{lm0}. But when extending the Fourier series to l_{max} = 10 covariance matrix of multidimensional regression has increased so much that the more great opposite bias appeared generating instability.

Figure 2.3 shows the affect variance of measuring of σ^2 on the errors of estimates for the best by accuracy and stability regression model with the l_{max} = 8.

Denoted as r_l the relative standard deviations of the average by modulus harmonics $|U|_l$ from the true values $|\hat{U}|_l$, depending on their degree of l were calculated by the total sample of size M=300.

When reducing the error of primary data the variance and bias of the measured harmonics are decreased in concert. Due to the optimal arrangement of observation points in reciprocal space of a polycrystal the effect of improving the accuracy of measurements becomes the greatest [83].

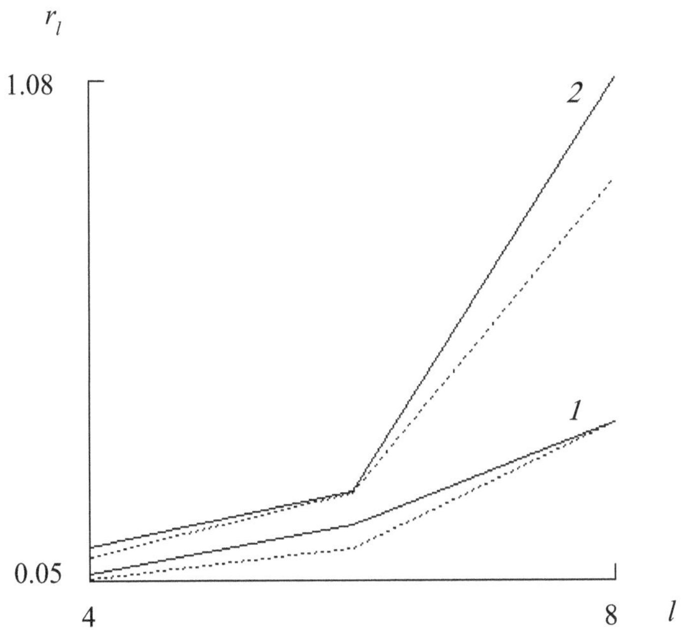

Fig. 2.3. Standard deviations of the estimated harmonics with respect to the true values under different variance of simulation measurements: (*1*) σ^2 = 0.01; (*2*) σ^2 = 0.04.
Dashed line for deviations from the sample means

Simulation experiments gave solid confirmation of the principles of determining the harmonics of the texture function as accurately as possible.

3. Tests of the methods of optimal measurement of crystallographic texture on low-carbon steel specimens. In criterion of optimality experiment for textural analysis of low-carbon steel will have to lay the reduction of experiment value at the expense of some loss of efficiency.

At bcc crystal lattice the sphere, covering the entire volume of the observations area in reciprocal space, is hard available to measuring. X-ray analysis of the reflection {222} greatly complicates the experiment. Abandoning too expensive information, we limit the observations area on sphere with reciprocal lattice site of {110}.

When limited the observations area the most preferred by largest radius r sphere is formed in the rotation of the radius-vector ⟨½ ½ 1⟩ of bcc reciprocal lattice. Here will to be scanning the reflection {112}.

Tables with points coordinates to D-optimum designs of experiment, designed for practical applications, are available in the automated system to research of crystallographic texture presented in Ch. 5.

Experimental studies of crystallographic texture of a sheet of low-carbon steel 08Yu with deformation extent of 72% were performed by D.A. Kozlov [38].

For preliminary harmonic analysis of pole figures X-ray measuring was carried out in Fe–K_α radiation [78, 80]. Mainly experiment, when being measured harmonics of the texture function, Co–K_α radiation used. There consistently is chosen the slits system and scan angle range of $\Delta 2\theta(r_{hkl})$. The average measuring time of one point is $\overline{t}_{\{110\}} = 40\ s$ and $\overline{t}_{\{112\}} = 100\ s$. As reference for determining the average intensities of reflections {110} and {112} is suitable specimen of steel 08Yu in weakly deformed state after hot rolling at 1200 C. Implementation the textural experiment corresponds to the technology [4].

The best by accuracy and stability model of the regression experiment to study thin metal sheet with cubic symmetry of the crystal, as follows from experience, contains the vector of harmonics of texture function with the highest degree of $l_{max} = 8$ whose dimension is $N = 13$. When bcc lattice of crystals, the number of observation points by D-optimal design of experiment is divided into $N_{\{110\}} = 10$ and $N_{\{112\}} = 3$.

Measured harmonics of the texture function of low-carbon steel are presented in Table 2.3. For chosen regression model an agreement with observations in statistical test by F-criterion is not rejected. According to t-test by available estimation of the covariance matrix of errors the harmonics are significant with probability $P = 0.99$.

Optimum experiment on accuracy of harmonics measurement gives and all pole figures that predicted from harmonics with the most reliability, as confirmed Fig. 2.4.

Table 2.3.

Measured harmonics of the texture function in the specimen of thin-sheet low-carbon steel

l	*Coefficients* U_{lm0}/U_{000}				
	$m = 0$	$m = 2$	$m = 4$	$m = 6$	$m = 8$
4	−0.9937 ±0.1461	−2.1024 ±0.0908	−1,6157 ±0.1272		
6	1.2221 ±0.0243	−0.6613 ±0.0151	0.2946 ±0.0148	−0.9322 ±0.0794	
8	−1.3656 ±0.0779	0.7659 ±0.0557	0.1962 ±0.0495	−1.1848 ±0.0923	0.8696 ±0.1259

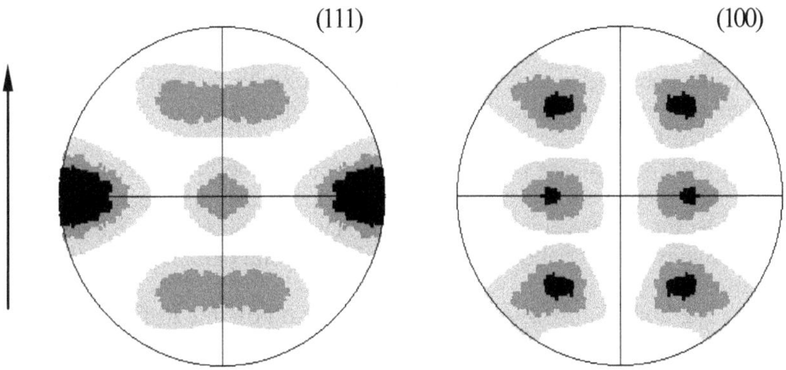

Fig. 2.4. The pole figures of the thin-sheet low-carbon steel when predicting by measured harmonics of the texture function (arrow shows the rolling direction)

Experimental verification provides the conclusion about the good quality of received harmonics of the texture function. Further in the Ch. 4 and Ch. 5 there is evidence of their successful use for determination of the textural components and study of the anisotropy of plasticity of metal sheet.

CHAPTER 3
DETERMINATION OF THE PARAMETERS OF THE TEXTURE FUNCTION BY MEASURED HARMONICS

There has been proved by experience that the statistical model of the crystal structure of deformed metals corresponds to reality. Method of the model estimation identifies the type of resolvable components of texture, their shape and quantitative ratio [66, 71].

§ 3.1. Statistical Estimation of a Mixed Distribution of the Orientations of Crystals

The essence of the problem is to present real crystallographic texture by components of the theoretical probability distribution of orientations.

1. Global search for the optimum texture function of investigated object. The number of separable peaks of probability density of orientations in general is limited by the dimension of the vector of observations **U**. When adequacy the model *f*(**g**,**B**) in form (1.1) with the least components number of *s* it has advantages in accuracy and stability of the parameter estimates of **B** [5].

The most significant property of estimates of the physical parameters is the statistical validity. Only then predictions are approaching to the truth with the improvement of the accuracy of original data. Requires at least achieve stability and validity in estimating the parameters of the texture function *f*(**g**,**B**).

The method of maximum likelihood make it possible to find in the allowable range of parameters **B** an estimate with the lowest determinant of the covariance matrix $\mathbf{V_B}$ satisfying the proposed requirements [105].

Maximizing the likelihood presumably normal data sampling with unknown covariance matrix is reduced to finding the minimum of the objective function [2]:

$$L(\mathbf{B}) = \frac{1}{2}\log\det\mathbf{M}(\mathbf{B}),$$

$$\mathbf{M}(\mathbf{B}) = \left[\mathbf{U} - \hat{\mathbf{U}}(\mathbf{B})\right]\left[\mathbf{U} - \hat{\mathbf{U}}(\mathbf{B})\right]^{t}.$$

Here, **M** is the matrix of the moments of deviations of the observed vector of normalized spherical harmonics **U** from the expected theoretical vector $\hat{\mathbf{U}}(\mathbf{B})$ (1.5).

To reduce the sensitivity of the objective function of the method of maximum likelihood to disturbance of the normal distribution of errors, it being minimized simultaneously with the stabilizing functional [95]:

$$F(\mathbf{B}) = L(\mathbf{B}) + \omega \,|\, \left\| \hat{\mathrm{U}}(\mathbf{B}) \right\|^2 \Rightarrow \min.$$

Using the coefficient of $0 < \omega \leq 1$, which should slowly decrease with approach to the optimum, there can choose the vector $\hat{\mathbf{U}}(\mathbf{B})$ with the smallest norm out of all which are consistent with the data **U**.

In global optimization are applied discrete and parametric methods.

Simple types of possible textural components $\{hkl\}\langle uvw\rangle$ is detected in the qualitative analysis of pole figures, which can always be constructed for the same observations of **U**. Search the textural components for a minimum of functional *F*(**B**) can be performed by exhausting all combinations the *s* types of permissible orientations that constitute the set of $\mathbf{\Pi} = (\mathbf{E}_1, \ldots, \mathbf{E}_\sigma)$.

Fourier coefficients of a discrete distribution of the texture highs are calculated originally for crystallographic orientation (001)[100] then are averaged over symmetric rotations of crystal. When choosing any type of orientations $\{hkl\}\langle uvw\rangle$, Fourier coefficients are transformed by turning crystal lattice to the nearest (*hkl*)[*uvw*].

Discrete search the most probable orientations $\mathbf{E}_\nu$ is performed with adjusted parts by weight μ_ν and the ordering parameters $\mathbf{K}_\nu$ ($\nu = 1, \ldots, s$).

It is useful to introduce the vector of normalized parameters **b** whose components have the same interval of acceptable values (0, 1):

$$\left.\begin{aligned} b_0^{(\nu)} &= \mu_\nu, \\ b_1^{(\nu)} &= \left(\mathrm{K}_1^{(\nu)} - \underline{\mathrm{K}}\right) / \left(\overline{\mathrm{K}} - \underline{\mathrm{K}}\right), \\ b_2^{(\nu)} &= \left(\mathrm{K}_2^{(\nu)} - \underline{\mathrm{K}}\right) / \left(\overline{\mathrm{K}} - \underline{\mathrm{K}}\right) \end{aligned}\right\} \quad (\nu = 1, \ldots, s).$$

Since the weight fractions of $(\mu_1, \ldots, \mu_s)$ are related, the vector **b** is of the dimension $3s - 1$.

Selection of interval $\left(\underline{\mathrm{K}}, \overline{\mathrm{K}}\right)$, which limits the minimum and maximum value of the order parameters, is provided the formula of the average cosine of angles of the normal random deviations of vectors from the expected direction: $\langle \cos\tau \rangle = \coth \mathrm{K} - \mathrm{K}^{-1}$ [19].

Value of $1 - \langle \cos\tau(\mathrm{K}) \rangle$ outside the range of $4 < \mathrm{K} < 60$ behaves asymptotically with $\underline{\mathrm{K}} \to 0$ and $\overline{\mathrm{K}} \to \infty$. If remain within the specified range, it is prevented the strong instability in estimating the parameters of $\mathbf{K}_\nu$.

The points in discrete searching of suitable parameters are the vertices of the hypercube, the total number of which is equal to 2^{3s}. The coordinates of the vertices this is vector-line of dimension of $3s$. Its elements are the boundaries of the interval of restrictions.

The lowest value of $F(\mathbf{B})$ is sought near the vertices of hypercube with random offsets to inside. Here, it is used an asymptotic representation of the extreme values in a sample of uniform distribution on interval [0, 1]:

$$\left[z/k ,\ \left(1 - z/k\right) \right];$$

k is a size of imaginary sample close of 10; z is a random variable with a distribution density $p(z) = e^{-z}$ [25].

Repeating the procedure of discrete optimization with added a randomness enable choose the best of several starting points of $\mathbf{B}^0$ to search the estimate of the structural parameters $\mathbf{b}^*$ under the found most probable orientations with coordinates $\mathbf{E}^*$.

The minimizing sequence of vectors $\mathbf{b}^{(i)}$, which are elements of $\mathbf{B}^{(i)}$, is calculated by formula of Newton. To approximate the inverse matrix of the second derivatives of the objective function of $\mathbf{H}^{-1}(\mathbf{B})$ that appears in the Newton formula it is used one of the methods of variable metric which is considered the most effective. Sequence of approximate matrices $\tilde{\mathbf{H}}^{-1}\left(\mathbf{B}^{(i)}\right)$ ($i = 0, 1, 2, \ldots$) converges to the exact $\mathbf{H}^{-1}\left(\mathbf{B}^*\right)$, and if agreed the model and data, it will be an approximation of the covariance matrix of the parameters $\mathbf{V}_\mathbf{B} \cong \mathbf{H}^{-1}\left(\mathbf{B}^*\right)$ [2].

Matrix of the moments of residual deviations $\mathbf{M}(\mathbf{B}^*)$ adjusted for bias by ζ gives an estimate of the covariance matrix of measurement, that is $\tilde{\mathbf{V}}_\mathrm{U} \cong \zeta \mathbf{M}\left(\mathbf{B}^*\right)$ [2]. Correction $\zeta = \left[1 - (6s - 1)/N\right]^{-1}$ takes into account $(6s - 1)$ fitting parameters relating to N equations of the model. The number of equations is equal to the dimension of the vector of observations $\mathbf{U}$. Parameters include Euler coordinates of the most probable orientations.

It is necessary to test the hypothesis that $\tilde{\mathbf{V}}_\mathrm{U}$ corresponds to the expected covariance matrix $\hat{\mathbf{V}}_\mathrm{U}$ of distribution of errors using estimated matrix $\mathbf{V}_\mathrm{U}$ from regression analysis of the measurements. Statistical criterion η to compare the $(N\times N)$ covariance matrices of multivariate samples of normal distribution was calculated in [36].

Only the diagonal elements of matrix $\tilde{\mathbf{V}}_\mathrm{U}$ can be estimated from one multidimensional observation $\mathbf{U}$, so taken approximate criterion

$$\eta \cong \sum_{k=1}^{N} \left[\log\left(v_{kk} / \tilde{v}_{kk} \right) + \tilde{v}_{kk} / v_{kk} \right] - N,$$

where $\tilde{v}_{kk}$ and v_{kk} respectively denote the diagonal elements of matrices $\tilde{\mathbf{V}}_{\mathrm{U}}$ and $\mathbf{V}_{\mathrm{U}}$. The asymptotic distribution of η is the χ^2 with the number of degrees of freedom equal to the number of unrelated elements of the covariance matrix precisely $q = N(N+1)/2$ [2].

Criterion η, assuming a normal distribution of errors and the large samples for estimation of matrices, in existing situation is applicable only as an indicative for rejecting obviously bad decisions when it repeatedly surpasses the critical value to the asymptotic distribution.

There being chosen a model with a well-defined parameters $\mathbf{B}^*$ which to be more than three times greater than their errors as estimated by covariance matrix $\mathbf{V}_{\mathbf{B}}$. Then at any law distribution of errors, the parameters are significant with probability of $P > 0.9$ (Chebyshev inequality [37]). Allowability of the parameters $\mathbf{B}^*$ qualitatively is controlled on pole figures predicted with model harmonics of $\hat{\mathbf{U}}\left(\mathbf{B}^*\right)$ compared to the forecast of measured harmonics of $\mathbf{U}$.

As an acceptable approximation to the true distribution of random orientations of crystals, it is taken an allowable decision of optimization problem, stably repeating within the errors. So it must be, if the decision satisfies the principle of maximum likelihood.

2. Representation of the orientations distribution density in a random aggregate of crystals. Constructed model is studied by making the probabilistic experiments. There is simulating test on specimen that showing a random distribution of orientations in ensembles of crystals.

Empirical distribution is a statistical analogue of the mixed distribution obtained by the convolution of functions from Eq. (1.2), (1.3):

$$f(\mathbf{g}, \mathbf{B}) = \sum_{\nu=1}^{S} \mu_\nu \left\langle Q\left(\mathbf{e}^{-1}\mathbf{g}, \mathbf{K}_\nu\right) \right\rangle_{\mathbf{E}_\nu}.$$

Angle brackets denote averaging over all ordered orientations of the same type appearing under rotations around the axes of symmetry of the crystal.

Let $\mathbf{g} = \mathbf{g}(\psi, \vartheta, \varphi)$ is the observed orientation of the crystal in the external coordinate system, coinciding with the axes of the rhombic symmetry of a specimen and $\mathbf{e} = \mathbf{g}(\psi_0, \vartheta_0, \varphi_0)$ is one of its expected orientations $(hkl)[uvw]$. Let us introduce the vector of normalized random deviations of crystallographic vectors relative to the most probable directions:

$$\mathbf{X}=\begin{bmatrix}X_1\\X_2\end{bmatrix}=\begin{bmatrix}2K_1\,\rho_{hkl}\\2K_2\,\rho_{uvw}\end{bmatrix},\quad \begin{cases}\rho_{hkl}=1-\cos(\vartheta-\vartheta_0),\\ \rho_{uvw}=1-\cos\big(\psi+\varphi-(\psi_0+\varphi_0)\big).\end{cases}$$

Independent random variables X_1 and X_2 follow approximately χ^2- distribution with two degrees of freedom [103].

Consequently, for each distribution of orientations with the vector of parameters **K**, there exists a vectorial random field $\boldsymbol{\rho}(\mathbf{X},\mathbf{K})$ with independent components of $\rho_{hkl}(X_1,K_1)$, $\rho_{uvw}(X_2,K_2)$. To sample functions of a random field of $(\boldsymbol{\rho}_1,\dots,\boldsymbol{\rho}_M)$ are made correspond the probability densities of $(Q(\mathbf{X}_1),\dots,Q(\mathbf{X}_M))$ in the form of converted expression from Eq. (1.3):

$$Q(\mathbf{X})=\left(\frac{K_1 e^{K_1}}{\operatorname{sh}K_1}\right)\left(\frac{e^{K_2}}{I_0(K_2)}\right)\left\langle e^{-\frac{1}{2}(X_1+X_2)}\right\rangle.$$

Values of the random variables X_1 and X_2, having a distribution density $p(X)=\frac{1}{2}e^{-\frac{X}{2}}$, are generated by the formula $X=-2\log\xi$ [11], where ξ is a random variable with a uniform distribution on the interval [0, 1].

The set of $(Q(\mathbf{X}_1),\dots,Q(\mathbf{X}_M))$ simulates the empirical distribution of random orientations of crystals. Two-dimensional graphic image of empirical distribution represents the shape of texture components.

§ 3.2. Example Identifying the Texture Components in a Thin Metal Sheet

Method of statistical estimation of the distribution of crystal orientations was tested on specimen of technically pure copper deformed at 90%. Original data are harmonics of the texture function which have been measured by D.A. Kozlov by optimal methods (§ 2.2 and § 2.3).

Using theoretical model of spherical harmonics which are measured, it is required determine parameters of the orientation ordering in the crystallographic texture of specimen.

For discrete search of optimal structure is taken a subset of the revealed orientations on neutron diffraction pole figure of copper with the deformation of 96.1% [29].

Being agreed with the observed harmonics **U** the allowable decisions of optimization problem $\mathbf{B}^*$ are considered as the measured values of the parameters of the orientations distribution $f(\mathbf{g},\mathbf{B})$.

Table 3.1.

Estimates of the components of crystallographic texture of a thin sheet of copper

Most probable orientation	*The weight fractions of texture components*	*Parameters of the orientations distribution*	
		$K_{\{hkl\}}$	$K_{\langle uvw\rangle}$
{112} ⟨111⟩	0.84 ± 0.01	6.2 ± 1.0	5.2 ± 1.0
{110} ⟨112⟩	0.16 ± 0.01	51.4 ± 5.2	50.1 ± 6.0

Table 3.2.

Validation of the texture function of the test specimen from measurement data

Indexes $l\ m\ n$	*Measured harmonics* U_{lmn}	*Predicted harmonics* $\hat{U}_{lmn}$	*Residual deviation*	*Prediction error*	*Measurement error*
4 0 0	−0.2243	−0.2635	0.0392	0.0732	0.0686
4 2 0	−0.0776	−0.0264	−0.0512	0.0677	0.0429
4 4 0	−0.7297	−0.7255	−0.0041	0.0634	0.0521
6 0 0	−0.4077	−0.2067	−0.2010	0.0319	0.0241
6 2 0	0.3001	0.1036	0.1965	0.0185	0.0109
6 4 0	−0.3033	−0.2506	−0.0527	0.0240	0.0104
6 6 0	−0.3119	−0.1803	−0.1316	0.0273	0.0193
8 0 0	−0.3499	0.3020	−0.6519	0.0955	0.0908
8 2 0	0.6558	0.0879	0.5679	0.0561	0.0553
8 4 0	−0.6482	−0.4870	−0.1612	0.0567	0.0350
8 6 0	1.3260	0.5029	0.8231	0.0607	0.0360
8 8 0	−0.1735	0.0746	−0.2481	0.1039	0.1037

Table 3.1 shows the average estimates of parameters over sample of measurements $\{\mathbf{B}_k^*\}$ ($k = 1, \ldots, 7$). Standard deviations of sample estimates are within the dispersion of their errors estimated by covariance matrix $\mathbf{V_B}$. The model harmonics of $\hat{\mathbf{U}}\left(\mathbf{B}^*\right)$ are agreed with the measured $\mathbf{U}$, that being checked by Table 3.2.

Figure 3.1 shows that model harmonics $\hat{\mathbf{U}}\left(\mathbf{B}^*\right)$ of the texture function of thin copper sheet reproduce the classical kind of pole figure (111) [29].

Such the same the most likely orientations of {112}⟨111⟩ and {110}⟨112⟩ which are identified by means of discrete optimization, are found when analyzing the pole densities calculated from the measured harmonics of the texture function $\mathbf{U}$ [84].

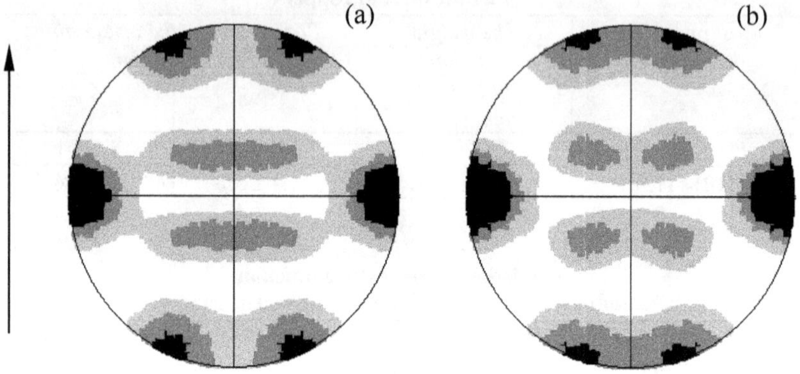

Fig. 3.1. Predicted the pole figure of (111) for specimen of copper (a) by the measured harmonics and (b) by model of the orientations distribution

Let $\mathfrak{R}$ is a joint pole density of any pair of orthogonal vectors, such that $\{HKL\} \parallel Z$, $\langle UVW \rangle \parallel X$. Sample to be formed of predicted pole densities for all feasible pairs with a simple indexes of $\{HKL\}\langle UVW \rangle$ by constructing a variation row of $\mathfrak{R}_1 < \mathfrak{R}_2 < \cdots < \mathfrak{R}_q$ (q is number of members) can be divided into statistically homogeneous groups based on the criterion of least significant differences [62]:

$$\left|\mathfrak{R}_{(\lambda)} - \mathfrak{R}_{(\lambda+1)}\right| < t_P \sqrt{\sigma^2_{(\lambda)} + \sigma^2_{(\lambda+1)}} \quad (\lambda = 1,\ 2,\ \ldots).$$

Here, there are involved the variances of sample members $\mathfrak{R}$, their estimates are calculated from the covariance matrix $\mathbf{V}_{\mathrm{U}}$ of measured harmonics $\mathbf{U}$, as well the percentage points for the t-distribution with N degrees of freedom according to the dimension of the vector $\mathbf{U}$. Specified confidence probability is $P = 0.99$.

If a jump is reliably identified on variation row, the type of the orientations density fluctuation is recognized in principle. Last element in variation row it is a possible fluctuations maximum.

Empirical distribution of orientations of the crystals in a specimen of a thin sheet of copper derived by simulation is shown in Fig. 3.2.

Graphical presentation of empirical distribution runs in a form of two-dimensional bar chart. The width of the columns corresponds to 95% ranges of uncertainty of orientations at existing variance of the estimates of parameters:

$$\left|\frac{\Delta\rho_{hkl}}{\rho_{hkl}}\right|_{\nu} = \left|\frac{\Delta K_1}{K_1}\right|_{\nu}, \quad \left|\frac{\Delta\rho_{uvw}}{\rho_{uvw}}\right|_{\nu} = \left|\frac{\Delta K_2}{K_2}\right|_{\nu}.$$

Height of the columns randomly varies in line with presumably normal errors of the weight fractions of the texture components of $\{\mu_\nu\}$. For graphical presentation sample size of 500 proved to be optimal when examined [84].

In Figure 3.2 the origin of Cartesian oblique coordinates of $(\rho_\alpha, \rho_\beta)$ is aligned with the crystallographic orientation of type $\{112\} \parallel Z$, $\langle 111 \rangle \parallel X$ (*Z* is normal to the specimen plane, *X* is rolling direction); $\rho_\alpha = 1 - \cos\alpha$, $\rho_\beta = 1 - \cos\beta$; α and β are angles of deviations about *X* and *Z* within 45°.

Confidence intervals of deviations of the ordered crystallographic vectors from the most probable directions, just as empirical distribution of the orientations are built on the basis of χ^2-distribution for normalized random variables of $\boldsymbol{\rho}$.

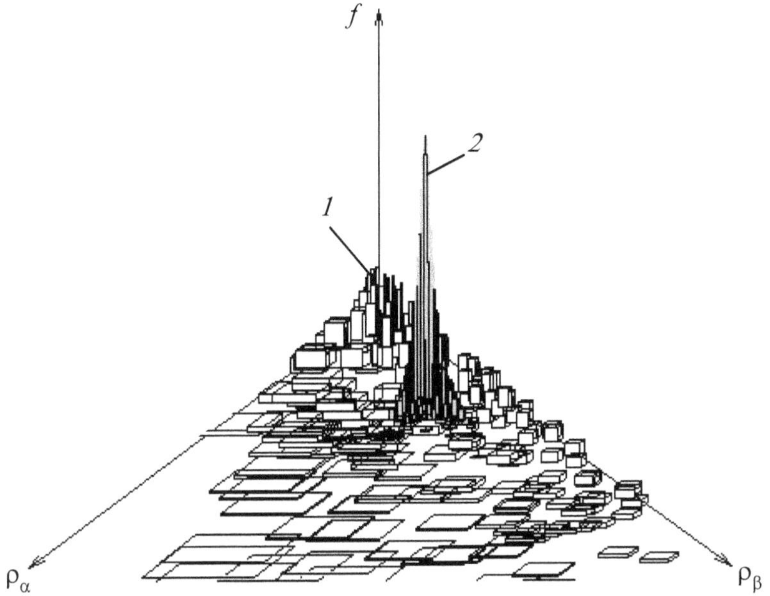

Fig. 3.2. Distribution of the orientations of a random ensemble of crystals in thin sheet of copper: (*1*) {112} ⟨111⟩; (*2*) {110} ⟨112⟩

Table 3.3.

Intervals of the angles of the crystallographic planes dispersion relative to plane of the copper sheet [deg]

{hkl}	*Confidence level*				
	0.50	*0.80*	*0.90*	*0.95*	*0.99*
{112}	27 ± 4	42 ± 7	51 ± 8	59 ± 9	75 ± 12
{110}	9 ± 1	14 ± 1	17 ± 2	20 ± 2	24 ± 2

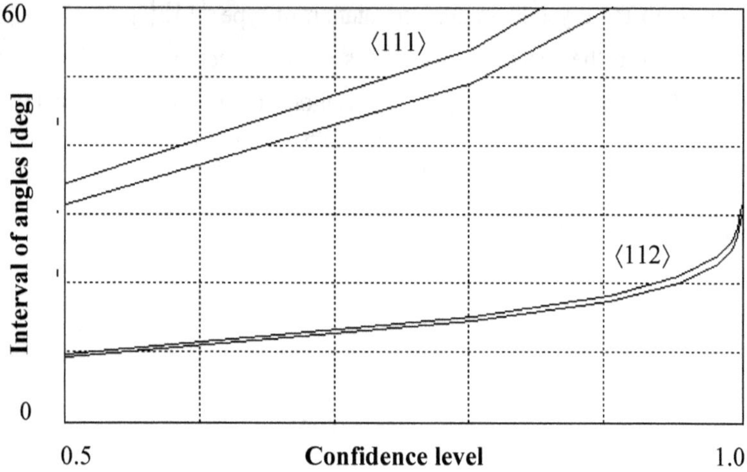

Fig. 3.3. Deviations of the crystallographic vectors ⟨*uvw*⟩ from rolling direction of the copper sheet. Reliability of the border width of the intervals of 95%

As examples for the identified texture of a specimen of copper the confidence intervals of the orientations dispersion relative to the plane of the sheet are shown in Table 3.3, and others in the itself plane of the sheet are shown graphically in Fig. 3.3.

Texture analysis using the method of separation of a mixed distribution of the orientations of crystals into components gives rise to the reliable representation of the orientation ordering under rhombic texture of metal sheet.

CHAPTER 4
STABILITY OF RHOMBIC TEXTURE AND CREATED BY IT ANISOTROPY

In a non-equilibrium polycrystalline system the plastic flow is interrelated with the kinetics of the orientations distribution of crystals. Movement of micro-plastic states of the system, caused by mechanical action, is regulated by the anisotropic hardening of differently oriented crystals and the relaxation of stresses in polycrystalline environment. Dynamic model of a deforming polycrystalline system is designed to study the macroscopic anisotropy inherent in crystallographic texture [76–77].

§ 4.1. Evolution of the Distribution of the Orientations of Crystals in the Process of Plastic Deformation

Kinetic equation for the distribution of the orientations of crystals reveals the natural qualities of crystallographic texture. In condition of existing symmetry the texture mobility is limited.

1. Probability distribution of the orientations of crystals in a non-equilibrium polycrystalline system. For any changes of probability distribution of crystallographic orientations $f(\mathbf{g})$ must be satisfied the condition of conservation of probability

$$\frac{\partial}{\partial t}\int_{\mathbf{G}} f(\mathbf{g})\,d\mathbf{g} = 0.$$

Consider the continuity equation for the probability density

$$\frac{\partial}{\partial t} f(\mathbf{r}) = -\operatorname{div}\mathbf{q}(\mathbf{r}),$$

where $\mathbf{r}$ is the rotating vector of three-dimensional Euclidean space, which is made correspond to the orientation of crystal $\mathbf{g}$ (rotation of vector can be described as a rotation of coordinate basis in which it is presented).

The current vector of probability

$$\mathbf{q}(\mathbf{r}) = f(\mathbf{r})\frac{d\mathbf{r}}{dt}$$

should be constructed so that the following equality holds

$$\frac{\partial}{\partial t}\int_V f(\mathbf{r})d\mathbf{r} = \int_V \frac{\partial}{\partial t} f(\mathbf{r})d\mathbf{r} = -\int_V \operatorname{div}\mathbf{q}(\mathbf{r})d\mathbf{r} = -\int_S \mathbf{q}(\mathbf{r})\cdot d\mathbf{s} = 0.$$

According to the equation of the transformation of vectors **r** when a continuous three-dimensional rotation with angular velocity $\boldsymbol{\omega}$, we have $\mathbf{q}(\mathbf{r}) = f(\mathbf{r})[\boldsymbol{\omega} \times \mathbf{r}]$ [37]. Vectorial element of surface *d*s is directed along **r**, so $(\mathbf{q}(\mathbf{r}) \cdot d\mathbf{s}) = 0$ that is the probability flux through bounding the volume *V* the surface *S* is missing, which proves the conservation of the orientations probability.

Thus, the rate of change of the probability density of the orientations of crystals is described by the equation

$$\frac{\partial}{\partial t} f(\mathbf{r}) = -[\mathbf{r} \times \Delta] \cdot (f(\mathbf{r}) \boldsymbol{\omega}(\mathbf{r})),$$

$$[\mathbf{r} \times \nabla] = \begin{bmatrix} y(\partial/\partial z) - z(\partial/\partial y) \\ z(\partial/\partial x) - x(\partial/\partial z) \\ x(\partial/\partial y) - y(\partial/\partial x) \end{bmatrix} = \begin{bmatrix} J_1 \\ J_2 \\ J_3 \end{bmatrix}.$$

Components of the vector [**r** × **∇**] are explicit differential expression of infinitesimal operators corresponding one-parameter subgroup of rotations around the coordinate axes (x, y, z) [16].

Presentation of the operators in the Euler angles of (ψ, ϑ, φ) of the rotation vector **r** follow [100]:

$$J_1 = \frac{1}{i}\left(\frac{\cos\varphi}{\sin\vartheta} \frac{\partial}{\partial\psi} + \sin\varphi \frac{\partial}{\partial\vartheta} + \operatorname{ctg}\vartheta \cos\varphi \frac{\partial}{\partial\varphi} \right),$$

$$J_2 = \frac{1}{i}\left(\frac{\sin\varphi}{\sin\vartheta} \frac{\partial}{\partial\psi} + \cos\varphi \frac{\partial}{\partial\vartheta} + \operatorname{ctg}\vartheta \sin\varphi \frac{\partial}{\partial\varphi} \right),$$

$$J_3 = \frac{1}{i}\left(\frac{\partial}{\partial\varphi} \right).$$

Introducing the vector operator of the $\mathbf{J} = J_1\mathbf{i} + J_2\mathbf{j} + J_3\mathbf{k}$, let us move to a more convenient form of the equation, to which must satisfy the probability distribution *f*(**g**):

$$\frac{\partial}{\partial t} f(\mathbf{g}) = -\mathbf{J} \cdot (f(\mathbf{g}) \boldsymbol{\omega}(\mathbf{g})). \tag{4.1}$$

This phenomenological equation describing the evolution of the macroscopic function of the orientations distribution[1], it is required relate with irreversible micro-processes proceeding in the system under study.

2. Phenomenological description of the micro-plastic states of a polycrystalline system. State of a deforming polycrystalline system can imagine as statistical ensemble of micro-plastic states of randomly oriented crystals. Microstates are defined by the values of the tensor the plastic strain rate $\acute{\varepsilon}_{ij}(\mathbf{g})$, or vector

$$\acute{\varepsilon}(\mathbf{g}) = \left(\acute{\varepsilon}_{11},\ \acute{\varepsilon}_{22},\ \acute{\varepsilon}_{33},\ \acute{\varepsilon}_{23},\ \acute{\varepsilon}_{31},\ \acute{\varepsilon}_{12}\right)^{\mathrm{t}}$$

(t-superscript denotes transpose). Form of a statistical ensemble of microstates fully is defined by probability distribution of the orientations of crystals, that is the texture function $f(\mathbf{g})$.

State of a polycrystalline system as a whole is characterized by tensor the rate of the macroscopic strain

$$\left\langle \acute{\varepsilon}_{ij} \right\rangle = \int_{\mathbf{G}} \acute{\varepsilon}_{ij}(\mathbf{g}) f(\mathbf{g}) d\mathbf{g}.$$

In averaging over the ensemble operates the current probability distribution of the crystallographic orientations. Tensor elements of $\left\langle \acute{\varepsilon}_{ij} \right\rangle$ are forming an integral observed $\left\langle \acute{\varepsilon} \right\rangle$.

Relations of micro-plastic states in the ensemble create micro-tension s_{ij}. The tensor macroscopic stress $\langle s_{ij} \rangle$, which generates movement in the system, there is a variable clearly connected with time.

For the elastic-plastic states of materials have been established the following fundamental ratios.

1. A linear relationship between the stresses and elastic deformations, and stresses and rates of inelastic deformations [45]:

$$s_{ij} = c_{ijkl}\varepsilon^{e}_{kl} + w_{ijkl}\acute{\varepsilon}^{p}_{kl}, \tag{4.2}$$

where s_{ij} is the sum of the elastic and "dissipative" stress tensor, ε^{e}_{kl} is tensor the elastic strain, $\acute{\varepsilon}^{p}_{kl}$ is tensor the plastic strain rate; c_{ijkl} and w_{ijkl} is tensors of elasticity and viscosity. (As to plastic component of the Eq. (4.2), it derives from the general expression for the dissipative function of deformable

---

[1] In [8] to describe the evolution of the texture function is adapted equation of hydrodynamics (rotation of crystals likened to the fluid flow) without determining the divergence in the system of curvilinear coordinates formed by Euler angles.

bodies, describing the internal friction, and only formally coincides with a viscous stress tensor in a fluid [45]).

2. Linear response to disturbance in ensemble of interrelated micro-plastic states [28]:

$$\left(\acute{s}_{ij}-\left\langle \acute{s}_{ij}\right\rangle\right)=-\chi_{ijkl}\left(\acute{\varepsilon}_{kl}^{p}-\left\langle \acute{\varepsilon}_{kl}^{p}\right\rangle\right), \quad (4.3)$$

where $\acute{s}_{ij}$ and $\left\langle \acute{s}_{ij}\right\rangle$ are tensors characterizing the rate of change of the local (non-uniform) and macroscopic (average) stress; χ_{ijkl} are relaxation coefficients of non-equilibrium system, depending on its internal structure. (Theory linear response asserts that a non-equilibrium state tends to move in equilibrium, exactly as in equilibrium any deviation from the mean tends on average to zero [58].)

Study of the mechanical properties of metals gave the following information [20]:

- elastic equilibrium when loading occurs almost instantly, therefore $\acute{\varepsilon}_{kl}^{e}=0$;
- changing in the elastic moduli during deformation is negligibly small, that is, $c_{ijkl}\approx \text{const}$;
- with static loading the effect of strain rate $\left(10^{-4}\leq \acute{\varepsilon}_{kl}^{p}\leq 10^{-1}\right)$ on the capacity to resist to (cold) deformation is insignificant, so that $\left(\partial w/\partial t\right)\approx\left(\partial w/\partial \varepsilon\right)\acute{\varepsilon}$ (w and $\acute{\varepsilon}$ are the fixed components of tensors w_{ijkl} and $\acute{\varepsilon}_{kl}^{p}$).

Equation of motion of the micro-plastic states get by combining the fundamental Eq. (4.2) and (4.3) under being in agreement with experience the assumptions:

$$\left.\begin{aligned}\frac{\partial}{\partial t}\acute{\varepsilon}_{ij}(\mathbf{g})&=h_{ijkl}(\mathbf{g})\left(\left\langle \acute{s}_{kl}\right\rangle-\chi_{klmn}(\mathbf{g})\Delta\acute{\varepsilon}_{mn}(\mathbf{g})\right),\\ \Delta\acute{\varepsilon}_{mn}(\mathbf{g})&=\acute{\varepsilon}_{mn}(\mathbf{g})-\left\langle \acute{\varepsilon}_{mn}\right\rangle.\end{aligned}\right\} \quad (4.4)$$

Here, $\Delta\acute{\varepsilon}_{mn}(\mathbf{g})$ are fluctuations of the field of strain rates over the orientations of crystals, h_{ijkl} is plasticity tensor as inverse to the tensor w_{ijkl} (index p omit, as further only plastic deformations are involved in the equations).

Shape of crystals is modeled by ellipsoids with principal axes ($a_1\geq a_2\geq a_3$) parallel to axes of the rhombic symmetry of the sample texture which have chosen as external coordinate system (X, Y, Z). Interrelation of the deformations of anisotropic crystal and its anisotropic polycrystalline environment (when not very much difference of elastic properties) is approximately represented by a model of the ellipsoidal inclusion undergoing the transformation in matrix [18].

Deformation local fluctuation $\left(\acute{\varepsilon}_{ij} - \left\langle \acute{\varepsilon}_{ij} \right\rangle\right) dt$ being constrained by environment creates the stress growth in the crystal

$$\left(\acute{s}_i - \left\langle \acute{s}_{ij} \right\rangle\right) dt = c_{ijkl} \left(\Gamma_{klmn} - \delta_{km}\delta_{ln}\right)\left(\acute{\varepsilon}_{mn} - \left\langle \acute{\varepsilon}_{mn} \right\rangle\right) dt,$$

hence it follows immediately that

$$\chi_{ijkl} = -c_{ijmn}(\Gamma_{mnkl} - \delta_{km}\delta_{ln}), \qquad (4.5)$$

and also causes a homogeneous rotation inside of it

$$\Omega_{ij}(t) dt = \Pi_{ijkl}\left(\acute{\varepsilon}_{kl} - \left\langle \acute{\varepsilon}_{kl} \right\rangle\right) dt, \qquad (4.6)$$

$$\mathbf{\Omega}(t) = \begin{bmatrix} 0 & -\omega_3 & \omega_2 \\ \omega_3 & 0 & -\omega_1 \\ -\omega_2 & \omega_1 & 0 \end{bmatrix},$$

that leads to a change in the distribution of crystallographic orientations.

Here, $\mathbf{\Omega}$ is the operator of infinitesimal rotations that related with angular velocity $\boldsymbol{\omega} = \omega_1\mathbf{i} + \omega_2\mathbf{j} + \omega_3\mathbf{k}$ at time t [37].

In the model are present the coefficients of constrained deformation Γ_{ijkl}, Π_{ijkl}, depending on the shape of the ellipsoid (δ_{ij} is Kronecker symbol). When shear deformations $\Gamma_{ijij} = \Gamma_{jiij}$, $0 \le 2\Gamma_{ijij} \le 1$ [18]. And as can be seen from the Eq. (4.5), the coefficient of elastic relaxation is $0 \le \chi_{ijkl} \le c_{ijkl}$. (Coefficient χ_{ijkl} reaches its maximum value of c_{ijkl}, when a crystal having been deformed becomes comparable to a thin plate so that $(a_1 \ge a_2 \gg a_3)$.)

Application of the model [18] means neglecting the disturbances of both stress and strain fields at the boundaries between crystals compared to fluctuations over the orientations.

3. Equation of the kinetics of the crystal orientation distribution in the Fourier representation. From the expression in Eq. (4.6), which determines the rotation of the crystal lattice in conditions of constrained deformation, follows the relation between the components of the angular velocity of rotation $\boldsymbol{\omega}$ and strain rate tensor:

$$2\omega_q = -e_{qij}\Omega_{ij} = -e_{qij}\Pi_{ijkl}\left(\acute{\varepsilon}_{kl} - \langle\acute{\varepsilon}_{kl}\rangle\right),$$

where e_{qij} is the Levi-Civita symbol, the non-zero elements of which $e_{123} = e_{231} = e_{312} = 1$; $e_{132} = e_{213} = e_{321} = -1$.

When assuming that coefficients Π_{ijkl} does not depend on **g** substitution of the expression for **ω** transforms the Eq. (4.1) to the following kinetic equation:

$$\frac{\partial}{\partial t} f(g) = \frac{1}{2} e_{qij}\Pi_{ijkl} J_q \left[f(g)\Delta\acute{\varepsilon}_{kl}(g)\right]. \quad (4.7)$$

For a description of real systems with random sizes of randomly oriented crystals naturally to take coefficients of the constrained deformation Π_{ijkl} for an average shape of crystals at the moment.

Let us represent the functions under the infinitesimal operator of J_q in the form of expansions on generalized spherical functions $T^{\mathbf{l}}(\mathbf{g})$:

$$f(\mathbf{g}) = \sum\nolimits_{\mathbf{l}} U^{\mathbf{l}}\, T^{\mathbf{l}}(\mathbf{g}),\ \ \Delta\acute{\varepsilon}_{ij}(\mathbf{g}) = \sum\nolimits_{\mathbf{l}} \Delta\acute{E}^{\mathbf{l}}_{ij}\, T^{\mathbf{l}}(\mathbf{g})$$

($\mathbf{l} = (l, m, n)$ it is vector of indices of the spherical harmonics of degree l).

Applying to both sides of Eq. (4.7) an integral transformation with the kernel $(2l+1)T^{\mathbf{l}}(\mathbf{g})^*$, where $T^{\mathbf{l}}(\mathbf{g})^*$ is a complex conjugate function, we obtain the equation for rate of changing the spherical harmonics of the texture function $f(\mathbf{g})$:

$$\left.\begin{aligned} &\frac{\partial}{\partial t}U^{\mathbf{l}} = \frac{1}{2} e_{qij}\Pi_{ijuv} \sum_{\mathbf{l}_2}\sum_{\mathbf{l}_2} U^{\mathbf{l}_1}\,\Delta\acute{E}^{\,\mathbf{l}_2}_{uv} \times \\ &\times (2l+1)\int_{\mathbf{G}} T^{\mathbf{l}}(\mathbf{g})^* J_q\left[T^{\mathbf{l}_1}(\mathbf{g})T^{\mathbf{l}_2}(\mathbf{g})\right]d\mathbf{g}. \end{aligned}\right\} \quad (4.8)$$

The product of generalized spherical functions in expression under the integral has expansion to the series of Clebsch - Gordan [100]:

$$T^{\mathbf{l}_1}(\mathbf{g})T^{\mathbf{l}_2}(\mathbf{g}) =$$

$$= \sum_{l'=|l_1-l_2|}^{l_1+l_2} C(l_1,l_2,l';m_1,m_2,m_1+m_2)\, C(l_1,l_2,l';n_1,n_2,n_1+n_2)\times$$

$$\times T^{l'}_{(m_1+m_2)(n_1+n_2)}(\mathbf{g}).$$

Therefore, in Eq. (4.8) there will be the sum of integrals of form

$$\Theta_q(\mathbf{l},\mathbf{l}') = (2l+1)\int\limits_{\mathbf{G}} T^{\mathbf{l}}(\mathbf{g})^* J_q\, T^{\mathbf{l}'}(\mathbf{g})\,d\mathbf{g}.$$

As a result of representations [100]:

$$J_1\, T^l_{mn}(\mathbf{g}) = i\frac{1}{2}\Big[\sqrt{(l-n)(l+n+1)}\,T^l_{m(n+1)}(\mathbf{g}) - \\ -\sqrt{(l+n)(l-n+1)}\,T^l_{m(n-1)}(\mathbf{g})\Big],$$

$$J_2\, T^l_{mn}(\mathbf{g}) = \frac{1}{2}\Big[\sqrt{(l-n)(l+n+1)}\,T^l_{m(n+1)}(\mathbf{g}) + \\ +\sqrt{(l+n)(l-n+1)}\,T^l_{m(n-1)}(\mathbf{g})\Big],$$

$$J_3\, T^l_{mn}(\mathbf{g}) = -nT^l_{mn}(\mathbf{g})$$

and orthonormality of generalized spherical functions [16]:

$$(2l+1)\int\limits_{\mathbf{G}} T^{\mathbf{l}}(\mathbf{g})^*\, T^{\mathbf{l}'}(\mathbf{g})\,d\mathbf{g} = \delta_{ll'}\delta_{mm'}\delta_{nn'}$$

the following components appear:

$$\Theta_1(\mathbf{l},\mathbf{l}') = i\delta_{ll'}\delta_{mm'}\frac{1}{2}\Big[\sqrt{(l'-n')(l'+n'+1)}\,\delta_{n(n'+1)} - \\ -\sqrt{(l'+n')(l'-n'+1)}\,\delta_{n(n'-1)}\Big],$$

$$\Theta_2(\mathbf{l},\mathbf{l}') = \delta_{ll'}\delta_{mm'}\frac{1}{2}\Big[\sqrt{(l'-n')(l'+n'+1)}\,\delta_{n(n'+1)} + \\ +\sqrt{(l'+n')(l'-n'+1)}\,\delta_{n(n'-1)}\Big],$$

$$\Theta_3(\mathbf{l},\mathbf{l}') = -n'\,\delta_{ll'}\delta_{mm'}\delta_{nn'},$$

which involve $m' = m_1 + m_2$; $n' = n_1 + n_2$ by definition Clebsch-Gordan coefficients.

In a system with cubic symmetry of crystals there are different from zero only harmonics U_{lmn} of order $n = 4k$ ($k = 0, 1, \ldots$) (Table 1.1), therefore in

Eq. (4.8) the terms with $\Theta_q(\mathbf{l}, \mathbf{l}')$, where q = (1, 2), under summation on indices $\mathbf{l}_1$ and $\mathbf{l}_2$ are turn into zero. For $q = 3$ the set of index values $i \neq j$ in the symbol e_{qij}, and hence $u \neq v$ in the coefficients Π_{ijuv} it is (1, 2), moreover $\Pi_{ijij} = -\Pi_{jiij}$ [18].

As a result, the equation of the kinetics of the crystal orientation distribution in the Fourier representation will be following:

$$\frac{\partial}{\partial t} U^{\mathbf{l}} = -n\left(2\Pi_{1212}\right) U^{\mathbf{l}} * \Delta \acute{E}^{\,\mathbf{l}}_{12}, \tag{4.9}$$

$$U^{\mathbf{l}} * \Delta \acute{E}^{\,\mathbf{l}}_{12} = \sum_{l_1=0}^{\infty} \sum_{m_1=-l_1}^{l_1} \sum_{n_1=-l_1}^{l_1} U_{l_1 m_1 n_1} \times$$

$$\times \sum_{l_2=\max(|m-m_1|,|n-n_1|)}^{\infty} C(l_1, l_2, l; m_1, m-m_1, m)\, C(l_1, l_2, l; n_1, n-n_1, n) \times$$

$$\times \Delta \acute{E}^{\,l_2 (m-m_1)(n-n_1)}_{12}.$$

Here, there is a convolution of the Fourier images in the space of generalized spherical functions, the formula of which has been deduced in [76].

4. Physical interpretation of the kinetic equation for the distribution of the orientations of crystals. As follows at once from the Fourier transformation of the kinetic equation the rate of changing of spherical harmonics $U^{\mathbf{l}} \equiv U_{lmn}$ of degree $l < 12$ is zero: when of cubic symmetry of crystals for $l < 12$ there is only one independent harmonic of the n-th order, precisely, of the order $n = 0$ (Table 1.1). However harmonics of the high degree characterize the local fluctuations of distribution of orientations of the crystals that related with the occurrence of short-range order.

The long-range order of orientations in a polycrystalline system gradually to create it is impossible. Ordering of orientations has to happen jointly that by jump would be originated crystallographic symmetry. Arising of the mechanically unstable structures of deformation (fragmentation of crystals during deformation of near-critical [107–109]) heralds the instantaneous transition into a new stable structural state [14].

Generalization of available information leads to the idea of the formation of crystallographic texture as a kinetic phase transition that creates a stable structure with its own symmetry properties.

Stability of the crystallographic texture is predicted by the kinetics Eq. (4.9):

1. Tension along the axes of rhombic symmetry does not affect the crystallographic texture.[1] Within the ellipsoidal crystallites oriented along the

---

[1] This result was obtained experimentally on a thin metal tape [33].

symmetry axes the changes in strains $(\acute{\varepsilon}_{11}dt,\ \acute{\varepsilon}_{22}dt,\ \acute{\varepsilon}_{33}dt)$ do not cause rotation.

2. Relative to the shears parallel to the deformation plane the crystallographic texture is invariant. Components that depend on the $(\acute{\varepsilon}_{31},\ \acute{\varepsilon}_{32})$ are disappearing due to the symmetry properties of crystals.

3. Shears parallel to the rolling direction $(\acute{\varepsilon}_{12})$ affect only the sharp part of the texture function.[1] Herein harmonics of the high degree are mobile, but their effect is locally.

From the analysis of the kinetics of the crystal orientation distribution it follows that the crystallographic texture does not undergo fundamental changes in the process of plastic deformation. Deformed texture saves the symmetry such that in an equilibrium.

§ 4.2. Predicted Anisotropy in a Current State of a Polycrystalline System

Solution to the dynamics equation of a non-equilibrium polycrystalline system simulates macroscopic deformation process revealing the properties of the material with the existing distribution of the orientations of crystals. Simulation model is tested on practice in analysis of the anisotropy of metal sheet.

1. Equation of the dynamics of a weakly non-equilibrium polycrystalline system. State of a non-equilibrium system at the current time uniquely is described by the equation of joint evolution of the strain rate of differently oriented crystals $\acute{\boldsymbol{\varepsilon}}(\mathbf{g})$, and of the orientations distribution density $f(\mathbf{g})$.

Physical parameters that depend on the orientations of crystals are own parameters of observed $\acute{\boldsymbol{\varepsilon}}(\mathbf{g})$ on the group of rotations of three-dimensional Euclidean space **G**. Parameters reacting to the state of non-equilibrium system, are controlling its dynamics.

For crystals of cubic symmetry there are three elastic and two plastic constants:

$$\mathbf{c} = (c_1,\ c_2,\ c_3)^{\mathrm{t}}\ \ (c_1 = \hat{c}_{1111},\ c_2 = \hat{c}_{1122},\ c_3 = \hat{c}_{1212}),$$

$$\mathbf{h} = (h_1,\ h_2)^{\mathrm{t}}\ \ (2h_1 = \hat{h}_{1111},\ {}^1\!/\!_4\, h_2 = \hat{h}_{1212},\ \hat{h}_{1122} = -{}^1\!/\!_2\, \hat{h}_{1111}),$$

it is meant the basis of crystal [53–54].

---

[1] On the experimental pole figures of thin metal tape after tension to failure in the directions 25° and 45° to the rolling direction there is observed the original symmetry of texture, only the shape of density maxima of the pole vectors changed [33].

In the transition to an external system of coordinates the tensors of elasticity and plasticity of cubic crystals are transformed according to the formulas

$$c_{ijkl}(\mathbf{g}) = c_2\delta_{ij}\delta_{kl} + c_3\left(\delta_{ik}\delta_{jl} + \delta_{il}\delta_{jk}\right) + \left(c_1 - c_2 - 2c_3\right)\rho_{ijkl}(\mathbf{g}),$$

$$h_{ijkl}(\mathbf{g}) = -h_1\delta_{ij}\delta_{kl} + \frac{1}{4}h_2\left(\delta_{ik}\delta_{jl} + \delta_{il}\delta_{jk}\right) + \left(3h_1 - \frac{1}{2}h_2\right)\rho_{ijkl}(\mathbf{g}),$$

$$\rho_{ijkl}(\mathbf{g}) = \sum_{n=1}^{3} A_{in}A_{jn}A_{kn}A_{ln};$$

ρ_{ijkl} is an operator of linear transformation of the fourth-rank tensors in rotation of the basis which is described by matrix **A(g)** [90].

The plastic potential of crystals of cubic symmetry is represented by a function [54]:

$$2\Theta = h_1\left[\left(s_{22} - s_{33}\right)^2 + \left(s_{33} - s_{11}\right)^2 + \left(s_{11} - s_{22}\right)^2\right] + h_2\left(s_{23}^2 + s_{31}^2 + s_{12}^2\right).$$

Condition of the stationary plastic flow being 2Θ = const. Means, that is increased a strength to plastic flow $\{s_{ij}\}$ can indicate on fall of the plasticity coefficients (h_1, h_2) what is related with the strain hardening of crystal.

In consequence of the condition of the yield, the vector of coefficients of plasticity of a straining crystal with orientation of **g** is defined by the equation

$$\frac{\partial}{\partial t}\mathbf{h} = -\mathbf{K}\left[3\mathbf{h} - \mathbf{h}^0\right]\acute{\mathbf{e}}(\mathbf{g}), \tag{4.10}$$

where $\acute{\mathbf{e}}(\mathbf{g}) = \left(\acute{e}_1, \acute{e}_2\right)^{\mathrm{t}}$ is the vector of strain rates in the crystallographic directions ⟨100⟩, ⟨111⟩.

Vectors constants $\mathbf{h}^0$ and **K** contain the elasticity limits of s_1^0, s_2^0 and coefficients of strain hardening $\left(\partial s_1/\partial e_1\right)$, $\left(\partial s_2/\partial e_2\right)$ by tension of a single crystal in the directions ⟨100⟩, ⟨111⟩:

$$\mathbf{h}^0 = M\begin{bmatrix}\frac{1}{2}\left(1/s_1^0\right)^2 \\ 3\left(1/s_2^0\right)^2\end{bmatrix}, \quad \mathbf{K} = \begin{bmatrix}\left(1/s_1^0\right)\left(\partial s_1/\partial e_1\right) \\ \left(1/s_2^0\right)\left(\partial s_2/\partial e_2\right)\end{bmatrix}.$$

(M is metric coefficient in units of measurement of s_{ij}). It is expected that in the study area of elastic-plastic states the deviations from initial plasticity coefficients of $\mathbf{h}^0$ are small so $\left(h_1^0 - h_1\right)/2h_1 \ll 1$, $\left(h_2^0 - h_2\right)/2h_2 \ll 1$.

When strain $\varepsilon > 10^{-3}$ the plastic flow is carried out already by multiple slip in crystals [42]. Therefore as evaluation of the deformation hardening coefficients in the field of elastic-plastic states is suitable the generalized data for the stage of rapid hardening of single crystals on stress-strain curve. According to these data it turns out that

$$\mathbf{K} = \begin{bmatrix} K_1 \\ K_2 \end{bmatrix} \approx \begin{bmatrix} 10^{-3} E_1 / s_1^0 \\ 10^{-3} E_2 / s_2^0 \end{bmatrix},$$

where E_1 and E_2 are the elastic moduli of crystal in tension on directions $\langle 100 \rangle$ and $\langle 111 \rangle$.

To determine the strain rate $\mathbf{\acute{e}}$ in a specified crystallographic direction when the orientation of the crystal of $\mathbf{g}$, it must be inverse transformation of observed $\acute{\varepsilon}_{ij}(\mathbf{g})$ to the crystallographic basis, precisely $\acute{e}_{kl} = A_{ik} A_{lj} \acute{\varepsilon}_{ij}$. The strain rate $\acute{e}_1$ along $\langle 100 \rangle$ can be found as average for all symmetric rotations of cube the component $\overline{\acute{e}_{11}}(\mathbf{g})$, and the strain rate $\acute{e}_2$ along $\langle 111 \rangle$ as a projection of $\overline{\acute{e}_{11}}(\mathbf{g})$ on this direction.

Since $\sum_{k=1}^{3} A_{ik} A_{kj} = \delta_{ij}$, $\sum_{q=1}^{3} \acute{\varepsilon}_q = 0$ (constancy of volume during plastic deformation), there is obtained the following expression:

$$\langle \acute{\varepsilon} \rangle = \sum_{l=0}^{\infty} (2l+1)^{-1} \sum_{m=-l}^{l} \sum_{n=-l}^{l} \mathbf{\acute{E}}_{lmn} U_{lmn}.$$

Onto the consistency micro-plastic deformation in polycrystalline environment are affected the coefficients of elastic relaxation $\chi_{ijkl}(\mathbf{g})$ (4.5). Poisson's ratio ν, required for the calculation included in $\chi_{ijkl}(\mathbf{g})$ coefficients of the constrained deformation Γ_{ijkl}, is evaluated by the isotropic part of the elastic moduli of cubic crystals c_{ijkl}^0 [48]:

$$\nu \approx c_{1122}^0 \Big/ \left(c_{1111}^0 + c_{1122}^0\right) = c_3 / 2\left(c_1 - c_2 + c_3\right).$$

This approach is consistent with the assumption of the same coefficients of the constrained deformation Γ_{ijkl}, Π_{ijkl} for all orientations of crystals in the current state of a system: uneven changing a shape of variously oriented crystals, in comparison to initial own non-uniformity can be neglected.

We introduce the vector of system states $\acute{\mathbf{E}} = \{\ \acute{\mathbf{E}}^{\mathbf{l}}\ \}$, which is the Fourier transform of the observed variables

$$\acute{\boldsymbol{\varepsilon}}(\mathbf{g}) = \sum\nolimits_{\mathbf{l}} \acute{\mathbf{E}}^{\mathbf{l}}\, T^{\mathbf{l}}(\mathbf{g})$$

in the a space of generalized spherical functions of $T^{\mathbf{l}}(\mathbf{g})$. Fluctuations $\Delta\acute{\varepsilon}_{ij}(\mathbf{g})$ will be represented by harmonics $\Delta\acute{E}^{\mathbf{l}}_{ij} = \acute{E}^{\mathbf{l}}_{ij} - \langle \acute{\varepsilon}_{ij} \rangle \delta(\mathbf{l})$, where $\acute{E}^{\mathbf{l}}_{ij}$ are elements of vector $\acute{\mathbf{E}}^{\mathbf{l}}$ constituted in the same way as $\acute{\boldsymbol{\varepsilon}}(\mathbf{g})$.

Interrelated with $\acute{\boldsymbol{\varepsilon}}(\mathbf{g})$ the coefficients of plasticity $\mathbf{h}$ are represented by vector of harmonics $\mathbf{H} = \{\ \mathbf{H}^{\mathbf{l}}\ \}$. Harmonics $\mathbf{H}^{\mathbf{l}} = \left(H_1^{\mathbf{l}}, H_2^{\mathbf{l}}\right)^{t}$ are definable provided conservation of a continuous medium:

$$\begin{cases} \operatorname{sign}\left(\acute{\varepsilon}_q(\mathbf{g})\right) = \operatorname{sign}\left\langle \acute{\varepsilon}_q \right\rangle \\ \quad (q = 1,\ 2,\ 3); \end{cases} \qquad \operatorname{sign}(x) = \begin{cases} 1, & x > 0, \\ 0, & x = 0, \\ -1, & x < 0. \end{cases}$$

Integral observable depending on the current distribution of orientations of crystals is calculated using spherical harmonics $\acute{\mathbf{E}}^{\mathbf{l}} \equiv \acute{\mathbf{E}}_{lmn}$, $U^{\mathbf{l}} \equiv U_{lmn}$:

$$\langle \acute{\boldsymbol{\varepsilon}} \rangle = \sum_{l=0}^{\infty} (2l+1)^{-1} \sum_{m=-l}^{l} \sum_{n=-l}^{l} \acute{\mathbf{E}}_{lmn}\, U_{lmn}.$$

Equation of dynamics of a non-equilibrium polycrystalline system, combining the Eq. (4.4) and (4.10), in the Fourier representation takes the form

$$\left.\begin{aligned} \frac{\partial}{\partial t}\acute{E}^{\mathbf{l}}_{ij} &= -\delta_{ij}\left[\langle \acute{s}_{\alpha\alpha} \rangle \delta(\mathbf{l}) + \lambda_{aauv}\Delta\acute{E}^{\mathbf{l}}_{uv}\right] * H_1^{\mathbf{l}} + \\ &+ \frac{1}{2}\left[\langle \acute{s}_{ij} \rangle \delta(\mathbf{l}) + \mu_{ijuv}\Delta\acute{E}^{\mathbf{l}}_{uv}\right] * H_2^{\mathbf{l}} + \\ &+ 3\left[\langle \acute{s}_{\alpha\beta} \rangle \delta(\mathbf{l}) + \lambda_{\alpha\beta uv}\Delta\acute{E}^{\mathbf{l}}_{uv}\right] * \left(H_1^{\mathbf{l}} * R^{\mathbf{l}}_{ij\alpha\beta}\right) - \\ &- \frac{1}{2}\left[\langle \acute{s}_{\alpha\beta} \rangle \delta(\mathbf{l}) + \mu_{\alpha\beta uv}\Delta\acute{E}^{\mathbf{l}}_{uv}\right] * \left(H_2^{\mathbf{l}} * R^{\mathbf{l}}_{ij\alpha\beta}\right), \\ \frac{\lambda_{\alpha\beta uv}}{(c_1 - c_2)} &= \frac{\mu_{\alpha\beta uv}}{2c_3} = \Gamma_{\alpha\beta uv} - \delta_{u\alpha}\delta_{v\beta}. \end{aligned}\right\} \qquad (4.11)$$

$$\left.\begin{aligned} \frac{\partial}{\partial t}\mathbf{H}^{\mathbf{l}} &= -\mathbf{B}\frac{1}{2}\sum_{q=1}^{3}\operatorname{sign}\left\langle\acute{\varepsilon}_q\right\rangle\left[3\mathbf{H}^{\mathbf{l}} - \mathbf{h}^0\delta(\mathbf{l})\right] * \acute{E}_q^{\mathbf{l}}, \\ \mathbf{B} &= \begin{bmatrix} B_1 \\ B_2 \end{bmatrix} = \begin{bmatrix} K_1 \\ \sqrt{1/3}\,K_2 \end{bmatrix}. \end{aligned}\right\} \quad (4.12)$$

Here, $R^{\mathbf{l}}_{ijkl}$ are spherical harmonics of the operator $\rho_{ijkl}(\mathbf{g})$ carrying out transformation of tensors $c_{ijkl}(\mathbf{g})$, $h_{ijkl}(\mathbf{g})$ when rotating the basis of cubic crystals.

In matrix $R^{\mathbf{l}}_{ijkl} \equiv \left\{R^{\mathbf{l}}_{pq}\right\}$ $\left(p = 1, \ldots, 6\ (q = 1, \ldots, 6)\right)$ for each $\mathbf{l}$ only are six independent elements, since $R^{lmn}_{23} = R^{lmn}_{44}$, $R^{lmn}_{31} = R^{lmn}_{55}$,, $R^{lmn}_{12} = R^{lmn}_{66}$. Coefficients R^{lmn}_{pq} of order $m = 6, 8, \ldots, l$; $n = 8, 12, \ldots, l$ as well as of odd degrees l are equal zero [77].

Expanded notation of the convolution of spherical harmonics which appears in the Fourier transformation of functions on a group $\mathbf{G}$ is shown in § 4.1. Delta function $\delta(\mathbf{I})$ under convolution leaves spherical harmonics unchanged.

The condition for convergence of the Fourier series for the observed strain rate of $\acute{\boldsymbol{\varepsilon}}(\mathbf{g})$:

$$\int_{\mathbf{G}} \acute{\varepsilon}_q(\mathbf{g})\acute{\varepsilon}_q(\mathbf{g})\,d\mathbf{g} = \sum_{l=0}^{\infty}(2l+1)^{-1}\sum_{m=-l}^{l}\sum_{n=-l}^{l}\acute{E}_q^{lmn}\acute{E}_q^{lmn} = \left\|\acute{\mathbf{E}}\right\|^2 < +\infty$$

satisfied if this is a dissipative system, as the norm of any vector that defines the state of a dissipative system is limited [22].

For the nonlinear system studied, where the increasing internal resistance reduces the external action the specified condition is satisfied.

Relative mean squares calculated with harmonics $\acute{\mathbf{E}}$ and $\mathbf{H}$:

$$\eta^2_{\acute{\boldsymbol{\varepsilon}}} = \int_{\mathbf{G}}\left|\acute{\boldsymbol{\varepsilon}}(\mathbf{g}) - \overline{\acute{\boldsymbol{\varepsilon}}(\mathbf{g})}\right|^2 d\mathbf{g} \Big/ \left|\overline{\acute{\boldsymbol{\varepsilon}}(\mathbf{g})}\right|^2 =$$

$$= \sum_{l=4}^{\infty}(2l+1)^{-1}\sum_{m=-l}^{l}\sum_{n=-l}^{l}\acute{E}_q^{lmn}\acute{E}_q^{lmn} \Big/ \acute{E}_q^0\acute{E}_q^0 \quad (q = 1, \ldots, 6),$$

$$\eta^2_{\mathbf{h}} = \int_{\mathbf{G}}\left|\mathbf{h}(\mathbf{g}) - \overline{\mathbf{h}(\mathbf{g})}\right|^2 d\mathbf{g} \Big/ \left|\overline{\mathbf{h}(\mathbf{g})}\right|^2 =$$

$$= \sum_{l=4}^{\infty}(2l+1)^{-1}\sum_{m=-l}^{l}\sum_{n=-l}^{l}H_p^{lmn}H_p^{lmn} \Big/ H_p^0H_p^0 \quad (p = 1, 2)$$

is a measure of the non-uniformity of plastic flow and strain hardening of variously oriented crystals.

2. Simulation of the deformation process in a polycrystalline system at a given mechanical action. Suppose that mechanical action on a polycrystalline system complies with the conditions of the proportional stationary loading so that $\mathbf{P}(t)=\mathbf{n}\xi t$ where **n** is the loading direction with rate ξ, which corresponds to unchangeable tensor of stress state of n_{ij}. Under these circumstances macroscopic stresses are uniform, and are increasing by a constant rate of $\langle \acute{s}_{ij} \rangle = \xi n_{ij}$.

Let us introduce the vector $\mathbf{Z}=\left(\acute{\mathbf{E}},\mathbf{U}\right)^{t}$ that completely defines the state of the system at time t. Here, $\acute{\mathbf{E}}=\acute{\mathbf{E}}\left(\mathbf{H}\right)$ is vector of spherical harmonics of the observed strain rates $\acute{\varepsilon}(\mathbf{g})$ of variously oriented crystals, **U** is the vector of spherical harmonics of the probability distribution of orientations $f(\mathbf{g})$. In the vector of control $\mathbf{H}\left(\acute{\mathbf{E}}\right)$ are represented the plastic properties of a non-uniform hardening crystals in a current state of the system.

Equation of a non-equilibrium polycrystalline system to be represented in a generalized form:

$$\frac{\partial}{\partial t}\mathbf{Z}=\mathbf{\Phi}\left(\mathbf{Z},\mathbf{H}\right),\quad \frac{\partial}{\partial t}\mathbf{H}=\mathbf{\Psi}\left(\mathbf{H},\mathbf{Z}\right). \tag{4.13}$$

Explicit form of evolutionary functions $\left(\mathbf{\Phi},\mathbf{\Psi}\right)$ has been determined of Eq. (4.11) – (4.12).

For self-regulating deformation process in a polycrystalline system to be described by the solution of Eq. (4.13) can be obtained only model in form differences, which establishes the dependence of the current state vector **Z** from its values in the previous moments of time (t_0, t_1, … , t_k, t_{k+1}, …):

$$\begin{cases} \mathbf{Z}_{k+1}=\mathbf{Z}_{k}+\mathrm{L}_{k}\left[\mathbf{\Phi}\left(\mathbf{Z}_{k},\mathbf{H}_{k}\right)\right]\Delta t_{k}, \\ \mathbf{H}_{k+1}=\mathbf{H}_{k}+\mathrm{L}_{k}\left[\mathbf{\Psi}\left(\mathbf{H}_{k},\mathbf{Z}_{k+1}\right)\right]\Delta t_{k}, \end{cases}$$

and L_k denotes the interpolation polynomial of finite differences method.

The magnitude of the step Δt_k which in start of solution is much less than for the subsequent process is automatically adjusted during operation of the model to support the stability of a finite-differences procedure [37].

Vector $\mathbf{Z}_0$ corresponds to the initial state with the known distribution of the orientations of crystals and the same of $\acute{\varepsilon}(\mathbf{g})=0$. Initial vector of control $\mathbf{H}_0=\mathbf{h}^0\,\delta(\mathbf{l})$ is set according to the elasticity limit (or yield strength) in tension of single crystal in the directions ⟨100⟩ and ⟨111⟩. External parameters, at which the system is subjected to mechanical action of $\mathbf{P}(t)$, are constant.

On the state vector $\mathbf{Z}_k$ ($k = 0, 1, \ldots$) affect not only the current coefficients of plasticity of variously oriented crystals of $\mathbf{h}(\mathbf{g})$ presented in the vector of control $\mathbf{H}_k$ but also the coefficients of constrained deformation in polycrystalline environment of $\mathbf{\Gamma} = \{\Gamma_{ijkl}\}$, $\mathbf{\Pi} = \{\Pi_{ijkl}\}$, depending on the shape of a deforming grain (crystallite).

Let us suppose that the ellipsoidal shape of grain with the orderly direction of principal axes (a_1, a_2, a_3) parallel to the axes of rhombic symmetry, existing in the system, under straining remains the same (when a resisting surrounding the grain does not rotate as a whole). To calculate the coefficients of the constrained deformation of the ellipsoidal grain it is required to construct the matrix of their shape for current sizes of $a_q(t) = a_q(0)\left[1 + \varepsilon_q(t)\right]$ $(q = 1, 2, 3)$:

$$\mathbf{D}(t) = \begin{bmatrix} 1 & (a_2/a_1)^2 & (a_3/a_1)^2 \\ (a_1/a_2)^2 & 1 & (a_3/a_2)^2 \\ (a_1/a_3)^2 & (a_2/a_3)^2 & 1 \end{bmatrix}.$$

Into real polycrystalline system the grain shape is not uniform. There is diapason of changing of the macro-deformation $\langle \varepsilon(t) \rangle$, where according to the matrix $\mathbf{D}(t)$ visible changes in the structure not occur, so that the current values of coefficients $\mathbf{\Gamma}(\mathbf{D}(t))$, $\mathbf{\Pi}(\mathbf{D}(t))$ are indistinguishable.

It is naturally to present dynamics of the constrained deformation coefficients, and therefore the relaxation coefficients that control interrelations within a system, as discrete transitions to new values at points $\langle \boldsymbol{\varepsilon} \rangle_\tau$ ($\tau = 1, 2, \ldots$), satisfying the condition

$$\tau = \sum_{q=1}^{3} \left| \langle \varepsilon_q \rangle \right| / (3 v_a),$$

where v_a is the observed coefficient of variation of random variables a_q ($q = 1, 2, 3$) with a log-normal distribution law $p_a(a_q)$ in the volume of material of V.

For matrix $\mathbf{D}(t)$ the current sizes of grain are assumed to be the average values

$$\langle \overline{a_q}(t) \rangle = \int a_q(0)\, p_a(a_q)\, da_q \int \left[1 + \varepsilon_q(t, \mathbf{g})\right] f(\mathbf{g})\, d\mathbf{g} =$$
$$= \overline{a_q}(0)\left[1 + \langle \varepsilon_q(t) \rangle\right].$$

Calculation of the original coefficients **Γ**, **Π** and of new, when the significant changes in the shape of average grain on the field of non-uniform grain sizes appear, is carried out according to formulas that can be derived from the decision of a problem of elasticity theory for an ellipsoidal inclusion in a matrix [18].

Let a grain in the shape of an ellipsoid oriented along the axes of rhombic symmetry is of sizes $a_1 > a_2 > a_3$. Constructing the matrix of grain shape **D** with the elements of $d_{ij} = a_j / a_i$ $(i, j = 1, 2, 3)$, we calculate the following coefficients:

$$\Gamma_{1111} = \alpha\left[1 - (A_2 - A_1)/(1 - d_{12}) - (A_3 - A_1)/(1 - d_{13}) + \beta A_1\right],$$

$$\Gamma_{1122} = \alpha\left[(A_2 - A_1)d_{12}/(1 - d_{12}) - \beta A_1\right],$$

$$\Gamma_{2211} = \alpha\left[(A_2 - A_1)/(1 - d_{12}) - \beta A_2\right],\ \Gamma_{2112} = \Gamma_{1212},$$

$$2\Gamma_{1212} = \alpha\left[(A_2 - A_1)(1 + d_{12})/(1 - d_{12}) + \beta(A_1 + A_2)\right],$$

$$2\Pi_{1212} = -2\Pi_{2112} = A_2 - A_1,\ \alpha = 1/2(1 - \nu),\ \beta = 1 - 2\nu,$$

$$A_1 + A_2 + A_3 = 1,$$

$$A_1 = (L_1 - L_2)d_{12}/(1 - d_{12}),\ A_3 = (1 - L_2)/(1 - d_{23}),$$

$$L_1 = nF(\theta, k),\ L_2 = nE(\theta, k),\ n = \sqrt{d_{23}/(1 - d_{13})},$$

$$\theta = \arcsin\sqrt{(1 - d_{13})},\ k = \sqrt{(1 - d_{12})/(1 - d_{13})},$$

$F(\theta,k)$ and $E(\theta,k)$ are normal elliptic integrals Legendre of the first and second kind (ν is Poisson's ratio).

Other non-zero coefficients of $\{\Gamma_{ijkl}, \Pi_{ijkl}\}$ are produced by cyclic permutation of the indices (1, 2, 3).

When expected stability of the plastic flow in the system with the strain hardening, solution of the equation of the model in the Fourier representation should converges to the solution of the original equation. An area of the state space, where there are structural conditions of instability, is outside region of the limitations of the model.

3. Verifying the deformation model by simulation the tensile testing of cold-worked sheet of low-carbon steel. By giving the parameters of external action that satisfied the conditions of static tensile testing, we can observe a macroscopic deformation process on a computer exactly as in the full-scale tests.

When thin metal sheet it is allowable a plane stress state with the tensor of a growth rate of macroscopic stresses in the form

$$\langle s_{ij} \rangle = \begin{bmatrix} \xi_1 \cos^2\varphi + \xi_2 \sin^2\varphi & (\xi_1 - \xi_2)\cos\varphi \sin\varphi & 0 \\ (\xi_1 - \xi_2)\cos\varphi \sin\varphi & \xi_1 \sin^2\varphi + \xi_2 \cos^2\varphi & 0 \\ 0 & 0 & 0 \end{bmatrix}.$$

Here, ξ_1 and ξ_2 are the rates of loading a sample on orthogonal axes, and φ is the angle between the direction of tension in the plane of sheet and the rolling direction (X-axis), or the rotation angle around the normal to plane of sheet (Z-axis); $\xi_2 = 0$ when uniaxial tension, $\xi_2 = \xi_1$ when symmetrical biaxial. Selected tension directions for subsequent trigonometric interpolation of data taken from the strain diagrams make up set of $\varphi = (0°, 30°, 45°, 60°, 90°)$.

Object characteristics under simulated mechanical tests are summarized in Table 4.1. These include the physical properties of crystals, precisely the constants of elasticity **c** and the yield strength $\mathbf{s}^0$ by tension of single crystal in the directions ⟨100⟩ and ⟨111⟩, as well as the shape ratios of average grain in a polycrystalline material (the variation coefficient of the grain sizes of v_a adopted equal to 0.1).

Table 4.1.

Physical and structural parameters of the object of simulating mechanical tests

Elastic constants **c** [GPa] [94]	*Yield strength* $\mathbf{s}^0$ [MPa] [50]	*Grain shape ratios* [38]
$c_1 = 230.1$	$s_1^0 = 83.0$	$a_1/a_3 = 7.5$
$c_2 = 134.6$	$s_2^0 = 132.0$	$a_2/a_3 = 5.0$
$c_3 = 116.6$		

Distribution of the orientations of crystals in thin-sheet low-carbon steel has been determined from the measured harmonics of the texture function (Table 2.3), Fig. 4.1 presents its form.

Measure of strain hardening of the material it is an increase the resistance to straining compared to an initial. The strength to the plastic flow onset $\sigma_0(\varphi)$ and the anisotropy coefficient of initial plastic strains $r_0(\varphi)$ (ratio of the changes of plastic strain in mutually perpendicular directions each of that is perpendicular to tension axis) are calculated by Hill formulas [27].

For calculation as the approximate plasticity tensor of the metal sheet with rhombic symmetry is adopted average, with existing distribution of the crystallographic orientations, tensor $\langle h_{ijkl}(\mathbf{g}) \rangle$ that calculated using the averaged operator of tensors transformation under rotation of the basis of crystal:

$$\langle \rho_{ijkl}(\mathbf{g}) \rangle \equiv \langle \rho_{pq}(\mathbf{g}) \rangle = R_{pq}^0 + \sum_{l=4}^{\infty} \sum_{m=0}^{l} \sum_{n=0}^{l} R_{pq}^{lmn} U_{lmn}.$$

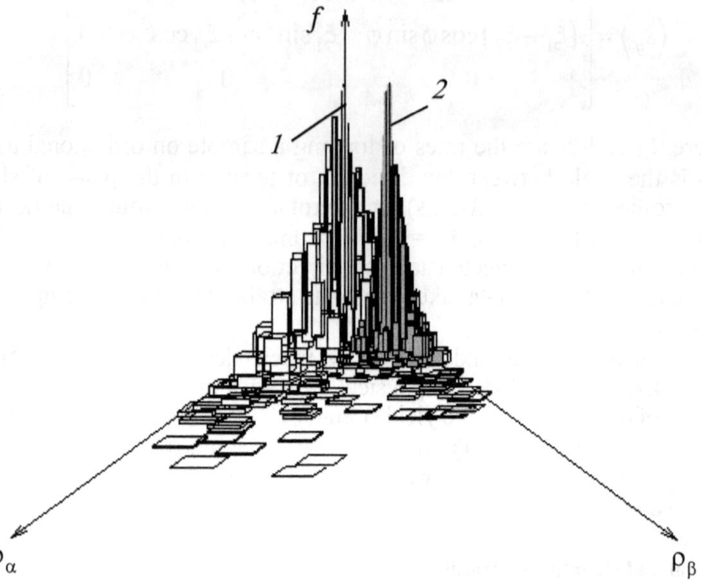

Fig. 4.1. Distribution of orientations of a sampled aggregate of crystals in the specimen of cold-worked low-carbon steel: (*1*) {112}⟨110⟩; (*2*) {111}⟨110⟩

Herein these are included the normalized spherical harmonics U_{lmn} of probability distribution of the orientations of crystals $f(\mathbf{g})$. The summands with $m = 0$ or $n = 0$ take weight of 0.5.

Table 4.2 presents the data on the changes of the tensile strength $\sigma(\varphi)$ and of the coefficient of normal plastic anisotropy $r(\varphi)$ in extreme points depending on the angle φ, which have been obtained by the simulated tests on uniaxial tension with a loading rate of 30 MPa·s^{-1}.

Table 4.2.

Anisotropy of strength and plasticity when different strain extent by the data of simulated tests of the metal sheet

Strain, ε [%]	*Tensile strength,* $\sigma(\varphi)$ [MPa]			*The plastic anisotropy coefficient,* $r(\varphi)$		
	$\varphi = 0°$	$\varphi = 45°$	$\varphi = 90°$	$\varphi = 0°$	$\varphi = 45°$	$\varphi = 90°$
0.0	114.70	102.98	109.12	1.1641	1.4225	0,9488
0.1	123,56	110.61	117.60	0.9796	1.4109	0.8181
10	205.30	181.43	195.60	0.9786	1.4061	0.8200

Figure 4.2 presents an overall picture of the macroscopic deformation process in different conditions of mechanical action (arrow shows the rolling direction).

Microinhomogeneity of deformation process which manifests itself in the fluctuations of plastic flow and of strain hardening of variously oriented crystals is shown in Fig. 4.3.

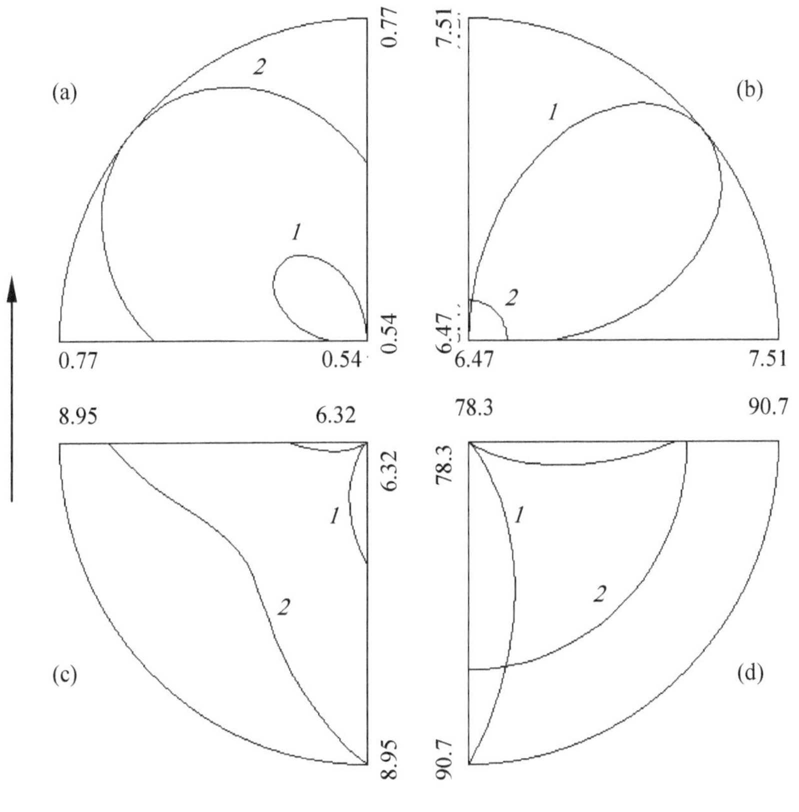

Fig. 4.2. Observed variables under simulation the tensile tests of metal sheet: (a), (b) the rate of plastic flow [s^{-1}]; (c), (d) the strain hardening [MPa]. Deformation of 0.1% [(a), (c)] with loading rate of (1) 20 and (2) 30 [$MPa{\cdot}s^{-1}$]. Deformation of 10% [(b), (d)] when tensile mode (1) uniaxial, and (2) biaxial with loading rate of 30 [$MPa{\cdot}s^{-1}$]

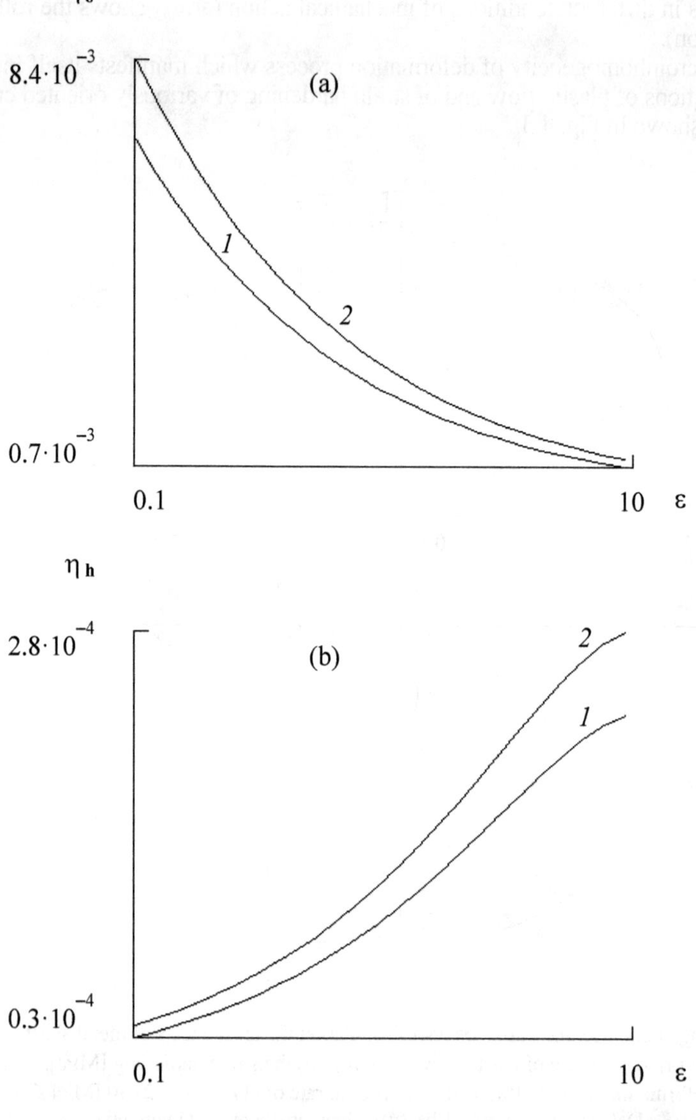

Fig. 4.3. The fluctuations evolution (a) of the vector of strain rate and (b) of the vector of plasticity coefficients over variously oriented crystals. Biaxial loading with rate of (1) 20 and (2) 30 [MPa·s^{-1}] (ε is strain extent [%])

Basic information from experiments on the simulation model is generalized in the following paragraphs:

1. Increase of the tensile strength of $\delta\sigma(\varphi)$ as well as decrease of the coefficient of normal plastic anisotropy of $\delta r(\varphi)$ is greater in the sheet tension direction where the stronger the strength to the plastic flow onset $\sigma_0(\varphi)$. Irregularity of changes according to directions during deformation, onto the form of dependency of $\sigma(\varphi)$ and $r(\varphi)$ does not have a significant effect. Initial anisotropy of mechanical properties of the metal sheet remains stable (Table 4.2).

2. The rate of macroscopic strain is in inverse dependence on the tension direction in the sheet plane as compared to the strain hardening $\delta\sigma(\varphi)$. The existing anisotropy is weakly sensitive to the rate of loading limited by conditions of static tests. With acceleration a loading the plastic flow and the strain hardening are increasing and becoming more non-uniform on crystal orientations (Fig. 4.2 and 4.3).

3. Fluctuations of the strain rate and the plasticity coefficients of variously oriented crystals during deformation undergo reverse changes. Development of process tends to equalize the plastic flow by increasing microinhomogeneity of strain hardening, it can be seen from the fact that $\eta_{\dot{\varepsilon}}$ is reduced while increasing $\eta_{\mathbf{h}}$ (Fig. 4.3).

The quality of approximation of the solution of the original equation of a non-equilibrium polycrystalline system can be verified at given different limits of the highest degree of spherical harmonics in infinite-dimensional state vector of **Z**. Observations confirm the stability of the solution when increasing dimension of the state vector.

Table 4.3 compares the data of models of different accuracy for the tension process in the direction of 45° to the rolling direction at loading rate of 30 $\mathrm{MPa{\cdot}s^{-1}}$.

Simulation model that takes into account the harmonics degree of $l \geq 12$ reveals the effects associated with the evolution of crystallographic texture.

When into micro-processes there being included a motion of the orientations of crystals a falling the coefficient of initial plastic anisotropy of r_0 proved to be more, as if the plastic flow would experienced an increasing strength of σ_0. At the beginning, a motion of the orientations increases the strain stress as it is seen of $\delta\sigma$ subsequently relaxes it.

Table 4.3.

Effect of inaccuracies of the simulation model on observational data

Strain extent, ε [%]	*Strain hardening,* $\delta\sigma$ [MPa]	*Plastic anisotropy coefficient, r*	*Spherical harmonics highest degree*
0.1	7.6345	1.4109	8
	7.7101	1.3948	12
10	78.4487	1.4061	8
	78.1061	1.3901	12

Confined mobility of crystallographic texture under deformation of the metal sheet can to be neglected in practice.

Analysis of the information received, indicates that modeling the deformation processes in the metal sheet gives a picture of the phenomenon, which is consistent with the physical representations. Simulation model serves as means of studying the elastic-plastic properties of the metal sheet in the automated system of research of crystallographic texture.

CHAPTER 5
AUTOMATED SYSTEM OF RESEARCH OF CRYSTALLOGRAPHIC TEXTURE

In the interactive software system that implements the optimal strategy of texture studies, an algorithmic complex performing the estimation, prediction and simulation is basic. Practical examples of the application of the system give knowledge on organization of information support, management and servicing [82, 84–85].

§ 5.1. Analysis of the Texture Measurements under Optimal Design of Experiment

Texture information obtained in the experiment is provided in the best estimate of the regression model of the observed intensity distribution of diffraction.

1. Entering and processing the primary experimental data. The automated system puts on researcher a choice of regression model experiment satisfying the condition of its accuracy and stability, that is the number of estimated regression coefficients should be the minimal required for an acceptable approximation of observation data (with the same data accuracy it is the smaller, the weaker the texture).

For selected regression model it is provided *D*-optimal design of experiment with the minimum required number of measurement points of $\{\mathbf{r}_j\}$ ($j = j(\mu,\nu)$; μ is number of spheres $\{hkl\}$, and ν is number of points at the μ-th sphere). The number of points N is equal to the dimension of the vector of regression coefficients, which is the vector of harmonics of the texture function of $\mathbf{U} = \{U_{lmn}\}$ with the given highest degree $l_{\max}$. Researcher can add points to the constructed *D*-optimal design of experiment (or set up another design).

Data are entered into the regression analysis program using the built-in editor with accompanying background information.

Vector of measurements of each point $\mathbf{r}_j$ ($j = 1, 2, \dots, N$) contains six elements, precisely the number of pulses accumulated onto a diffraction line in four symmetrical positions of the sample and to background on the edges of line. The total number of rows is equal to N multiplied by the number of the samples.

A measurement time of diffraction line and background the program requests having opened the input windows. The average intensities on spheres of $\{hkl\}$ presented in the observation area, which are required to the normalization, are entered using the reference special window.

Service package of programs provides the control and correction of errors in data entry, as well as the security of the system from incorrect actions and the recovery of the operating mode. Saving of information required to continue work in case of forced or abnormal abort is performed automatically.

Under primary data processing the normalized intensities measured from a background level are calculated and then are averaged over four repeated measurements of each j-th point. Sample variance of repeated measurements s^2 is estimated by the paired differences of the average intensities $\overline{J_j}$ ($j = 1, 2, \ldots, N$) for the minimum required two samples (§ 2.2).

The processed results of measurements make up a vector **J** of dimension $2N$. After estimating the N coefficients of the regression model, the N observations will remain to test for agreement the model with the data.

Documented data package may be copied or printed.

2. Optimal estimation of harmonics of the texture function. For infinite-dimensional true vector of harmonics of the texture function the best estimate by the regression model of observed scattering intensity on a polycrystal will be vector of **U** with a minimum norm $\|\mathbf{U}\|^2$ (§ 2.2 и § 2.3).

In the automated system of research it is used the procedure of a randomization of data sample that enable to estimate the covariance matrix $\mathbf{V_U}$ and prevent a bias in estimate of **U** when a little repeated measurements.

Analysis of the experiment is performed in the following sequence [15]:

1. From data being combined over the two samples is extracted a large number pairs of data samples with a random set of the same number of repeated measurements at each point, as in the original sample.

2. According to obtained the M pairs of random samples there is evaluated the set of optimal regression estimates $\left(\mathbf{U}_1^*, \ldots, \mathbf{U}_M^*\right)$.

3. As a result there are computed sample mean of estimates $\overline{\mathbf{U}^*}$ with covariance matrix $\mathbf{V_U}$.

Accuracy of the approximation of observations and reliability of the harmonics estimate are tested by statistical criteria.

When verifying the agreement of the chosen regression model with the measurements data there is considered the critical value F-test with the (N,N) degrees of freedom for confidence probability of $P = 0.95$. The significance of the regression coefficients is checked by t-test with the N degrees of freedom for confidence probability of $P = 0.99$.

To check that estimates are uncorrelated it is used approximate criterion by Fisher:

$$\eta = \sqrt{(M-3)/2}\ \ln\frac{1+\rho}{1-\rho} > |\xi|_P,$$

where ρ is sample correlation coefficient, and $|\xi|_P$ is the critical level of the standard normal distribution for a given confidence probability of P [31].

The assumption of a diagonal covariance matrix $\mathbf{V_U}$ will be rejected when the correlations are significant with the reliability of $P \geq 0.99$.

The serving complex provides the assistance to researcher in situations where the statistical inferences on the estimation of the obtained data indicate a failure; either model is inadequate by observations, or all of its parameters within the errors.

If the constructed regression equation under statistical tests is not rejected, interactive software system organizes all further studies using the vector of the measured harmonics of **U** and its covariance matrix $\mathbf{V}_U$.

§ 5.2. Identification of Crystallographic Texture by Data of its Harmonic Analysis

With the mathematical models is carried out the most accurate and complete parametric and graphic representation of the texture of a specimen.

1. Symmetry and dispersion of crystallographic texture. All the traditional ways of representing the crystallographic texture is performed by predicting fluctuations of the diffraction intensity on spheres in the reciprocal space of a polycrystal.

The measured harmonics of the texture function $\{U_{lm0}\}$ are used for approximate calculation of the intensities $J(r,\psi,\vartheta)$ on the sphere $\{hkl\}$ of radius $r = \sqrt{h^2 + k^2 + l^2}$ at points with spherical coordinates ϑ, ψ by summing of the Fourier series (2.4). Summation of the series is stable, when each Fourier coefficient U_{lm0} has a stabilizing factor, which takes account of measurement error σ_{lm} [95]:

$$\lambda_{lm} = 1 - \left(\sigma_{lm}^2 / |U_{lm0}|^2\right).$$

By the calculations there are constructed routine pole figures and stereographic projections of crystallographic vectors oriented as along the normal to the sheet plane and the direction of rolling. Graphic images can be recorded on the disc to compare with those that will be obtained after determining the textural components, for example, as in Fig. 3.2.

The quantitative information contained in harmonics of the texture function can be presented under convolution as two texture generalized parameters, which are changed within the interval [0, 1]. Of these, s_{hkl} is a deviations measure of all crystallographic planes $\{hkl\}$ relative to the plane of the sheet, and s_{uvw} is dispersion measure of all crystallographic directions $\langle uvw \rangle$, lying in the plane of the sheet (§ 1.2).

Product $(s_{hkl} \cdot s_{uvw})$ there is degree of approach to a random distribution of orientations. The ratio of (s_{uvw} / s_{hkl}) serves to technological research as an indicator of controlled isotropy in the plane of the sheet under anisotropy across the sheet.

In determining the generalized parameters of dispersion s_{hkl} and s_{uvw} there are calculated sum of squares of the measured harmonics U_{lm0} with the stabilizing factors λ_{lm}. Errors of parameters s_{hkl} and s_{uvw} are estimated with covariance matrix of measurements $\mathbf{V}_U$.

Sensitivity of the generalized parameters of the orientations dispersion on a change of the crystallographic texture has been revealed in studies of low-carbon steel made by D.A. Kozlov [38].

2. Parameters of the texture components and the orientations distribution in a random sampling of crystals. Confidence intervals of dispersion of the ordering crystallographic vectors can be constructed after the separation of the texture into components. The number of recognizable components is limited by dimension of the observation vector of **U** with covariance matrix of errors $\mathbf{V}_U$ (Ch. 3).

Determination of parameters of a mixed probability distribution of the orientations of crystals to first is performed in fully automatic mode. The type of the most probable orientations can be detected on the set of predicted densities of crystallographic vectors in the directions of the axes of symmetry of a sample, if the observations data are quite accurate.

Allowability of the model of a mixed distribution of the orientations should be verified on pole figures and the stereographic projections. After reviewing the results achieved during global optimization, the researcher can continue the search for an optimal structure, specifying the expected types of texture components.

The model corresponding to data with well-defined parameters is used for simulating the probabilistic experiments, which display distribution of the orientations in random aggregates of crystals. In graphical representation of sampled distributions it is distinguished the form of the texture components. Oblique Cartesian coordinate system with the origin in a maximum of the interchanging textural components creates a three-dimensional image.

Repeating the probabilistic experiments on a computer multiply, it is possible to observe not only the form of the distribution of orientations, but also fluctuations of distribution on ensembles of grains, which are the source of the dispersion of primary measurements, even with negligibly small fluctuations of grains number in ensemble.

Empirical distributions of orientations of crystals in test specimens of copper and low-carbon steel, that being extracted by statistical simulation are presented in Fig. 3.1 and Fig. 4.1.

Tables and charts of confidence intervals of dispersion of the crystallographic vectors relative to expected directions quantitatively characterize the orientational order. Graphical display of the dispersion intervals it can be seen on the example of the texture of copper (Fig. 3.3).

The type of identified the most probable orientations, their weight fractions and the degree of ordering give the complete information on the crystallographic texture.

3. Analysis of textural transformations in low-carbon steel using the automated system of research. Optimum experiment revealed important qualities of the textural transformation in thin-sheet low-carbon steel during recrystallization [67].

Measurements of crystallographic texture of low-carbon steel of 08Yu in cold-worked by 72% and the recrystallized state using fast heating were performed by D.A. Kozlov. To optimal regression experiment the highest degree of harmonics of the texture function is $l_{max} = 8$ (§ 2.3).

Figure 5.1 shows the stereographic projections with the predicted levels of density of the orientations of crystallographic vectors along normal to the plane of the sheet and direction of rolling.

Measured harmonics show a significant increase dispersion of the crystallographic vectors in the sheet plane after recrystallization that affected the generalized parameters of texture dispersion presented in Table 5.1.

Optimization of the model of a mixed distribution of the orientations of crystals leads to the conclusion that in thin-sheet low-carbon steel during recrystallization under rapid heating the rolling texture is reprodused, with two reliably identifiable components of {111}⟨110⟩ and {112}⟨110⟩. Parameters of the orientations distribution as in deformed and in recrystallized specimen of low-carbon steel are presented in Table 5.2.

Constructed texture functions consistent to the original data are used for statistical modeling the orientations distribution into random aggregates of crystals from test specimens. Empirical distributions giving representation on change of the textural components shape as a result of recrystallization are shown in Fig. 5.2.

Table 5.1.

Generalized parameters of texture dispersion in thin-sheet low-carbon steel

Process forming texture	*Deviations from the sheet plane by* s_{hkl}	*Deviations from the rolling axis by* s_{uvw}
Cold rolling with reduction rate of 72%	0.7526 ± 0.0190	0.1823 ± 0.0074
Recrystallization under rapid annealing	0.8451 ± 0.0282	0.3642 ± 0.0226

Table 5.2.

Estimates of the components of the crystallographic texture of thin-sheet low-carbon steel

Sign	*Most probable orientations*	*Weights of the texture components*	*Parameters of the orientational order*	
			$K_{\{hkl\}}$	$K_{\langle uvw\rangle}$
(a)	{112} ⟨110⟩	0.753 ± 0.004	20.3 ± 0.6	14.3 ± 0.4
	{111} ⟨110⟩	0.247 ± 0.004	42.3 ± 1.3	44.0 ± 1.7
(b)	{112} ⟨110⟩	0.434 ± 0.008	23.3 ± 1.3	19.7 ± 0.5
	{111} ⟨110⟩	0.566 ± 0.008	23.1 ± 1.8	7.2 ± 0.2

(a) cold-worked state; (b) recrystallized state.

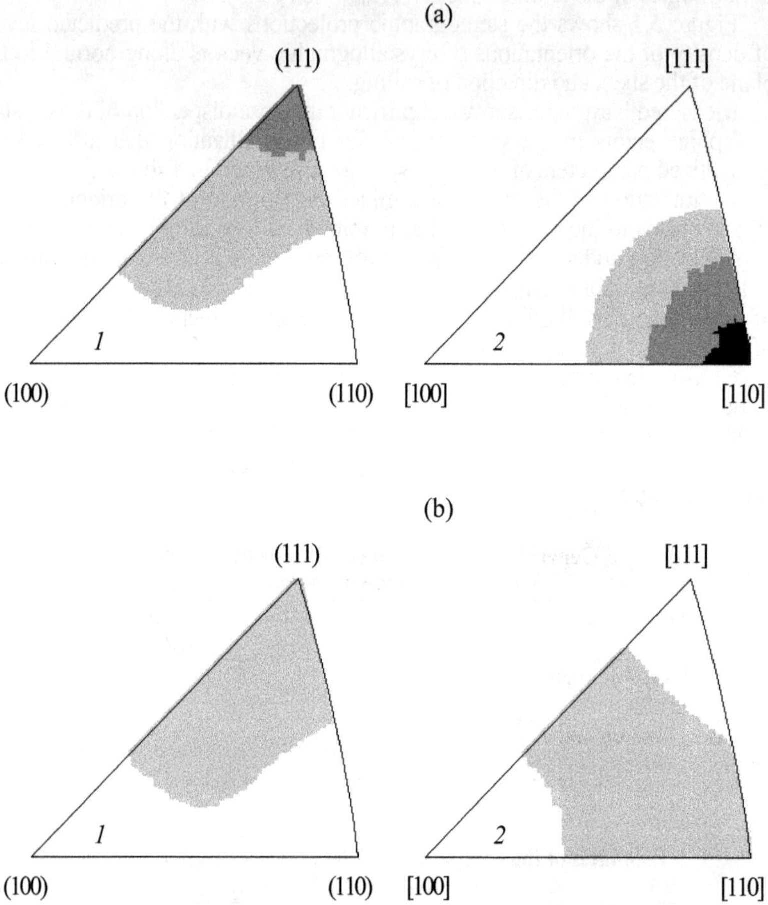

Fig. 5.1. Density levels of the crystallographic orientations in a thin sheet of low-carbon steel: (*1*) (*hkl*) at plane of the sheet, (*2*) [*uvw*] along the rolling direction; for specimen in (a) deformed state and (b) recrystallized state

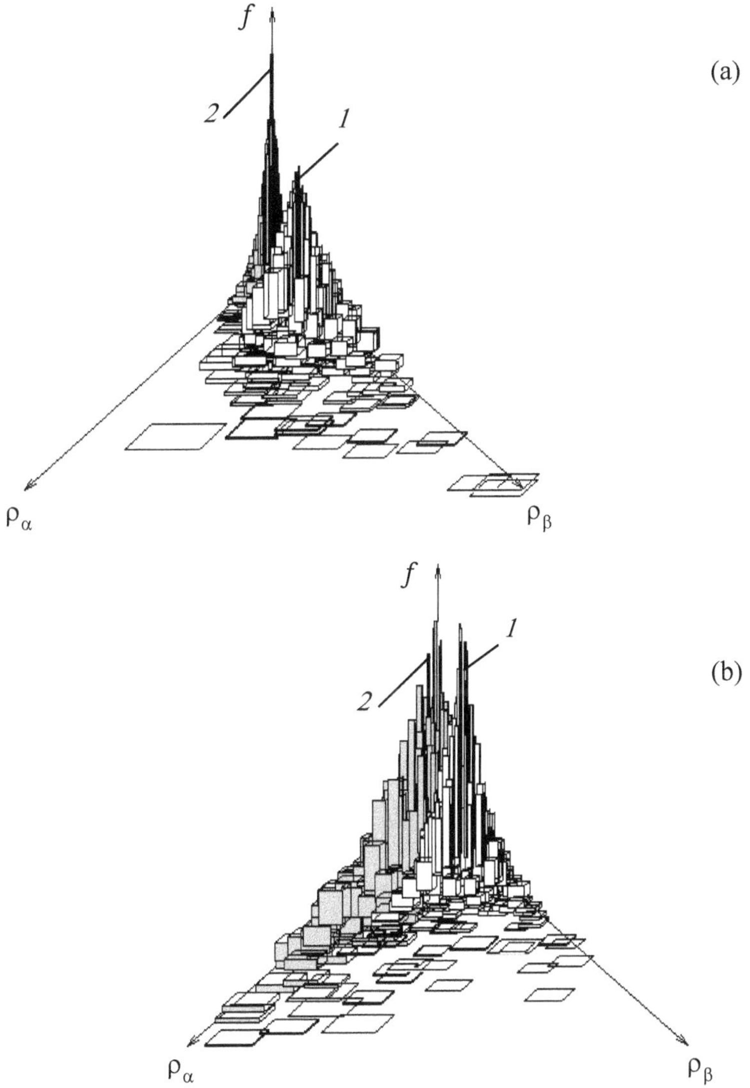

Fig. 5.2. A random distribution of the orientations of the sampled aggregates of crystals in thin-sheet low-carbon steel: (a) cold-worked state; (b) recrystallized state; (*1*) {112}⟨110⟩, and (*2*) {111}⟨110⟩

Table 5.3.

Maximum angle deviation from the rolling axis in the sharp texture component of low-carbon steel [deg]

States	*Confidence probability*				
	0.50	*0.80*	*0.90*	*0.95*	*0.99*
cold-worked	9–10	13–16	16–19	18–21	23–26
recrystallized	23–25	35–39	42–47	48–54	61–69

Width of the intervals border with confidence level of 95 %

A revealed broadening of texture maximum of {111}⟨110⟩ explains the preferential growth of generalized parameter of the crystallographic vectors dispersion in the plane of the sheet, that is, s_{uvw}. There is occurred disordering the vectors of ⟨110⟩ relative to rolling direction and so the interval of deviation angles has increased about 2.5 times (Table 5.3).

It was found that when rapid recrystallization of a thin metal sheet dominates growth of centers oriented by type of the most perfect textural component. Expanding in the plane of the sheet, the texture maximum of {111}⟨110⟩ gets doubled in the weight.

Quantitative transformation of the texture of the thin sheet of low-carbon steel discovers that crystallographic orientation with a high degree of order in cold-worked state is the highly active during recrystallization.

§ 5.3. Studying the Anisotropy of Strength and Plasticity of Metal Sheet by Measured Texture

When simulation of plastic deformation using the information on distribution of the orientations of crystals there is achieved so depth of study of the material that unavailable for full-scale tests.

1. Prediction of the initial anisotropy. Effective modules of elasticity and plasticity of metal sheet are calculated by averaging the modules of crystal with known distribution of its orientations. As a result of averaging the transition from cubic crystal to a polycrystalline system with rhombic symmetry occurs.

To calculation these are required the elastic constants of single crystal of c_{11}, c_{12}, c_{44}, and the elasticity limits (or yield strength) of s_1, s_2 in tension of single crystal on the directions of ⟨100⟩, ⟨111⟩. The program opens the input windows for the data.

Averaging the tensors of elasticity and plasticity of crystals is performed in the Fourier representation, using the measured vector $\{U_{lmn}\}$ of harmonics of the texture function and the matrix of spherical harmonics $\left\{R_{pq}^{lmn}\right\}$ of the transformation operator of the fourth-rank tensors when rotating the basis, in which they are defined (§ 4.2). The matrix $\left\{R_{pq}^{lmn}\right\}$ is contained in an auxiliary file involved to the automated system for studying the crystallographic texture along with the program of its calculation.

Effective elasticity coefficients of a polycrystal are calculated by the approximate formula [51]:

$$C^* = \frac{1}{2}\left[\langle c\rangle + \langle c^{-1}\rangle^{-1}\right].$$

Detailed calculation of the tensor of effective modules of elasticity and estimation of their variances by the covariance matrix of errors of the measured harmonics of $\mathbf{V}_U$ are described in [81].

Having the tensor of effective modules of elasticity of metal sheet, it is possible to construct curves of technical elastic modulus in function of the angle of rotation in the planes of rhombic symmetry of a sheet, perpendicular to the coordinate axes (X, Y, Z) [47].

As a tensor of effective modules plasticity of metal sheet it is adopted the average tensor for a polycrystalline sample with an existing the orientations probabilities distribution. Using an approximate tensor there are calculated the elasticity limits (or yield strength) and the coefficients of the normal plastic anisotropy for different directions in the planes of symmetry of metal sheet [27].

Example predicted from the measured harmonics of the texture function of the elasticity limit anisotropy of thin sheet of low-carbon steel 08Yu with deformation of 72% is presented in Fig. 5.3.

Comparison of the 95-percent confidence interval of forecasts with available measurements of maximum rise of the elasticity limit of iron after rolling for the strain extent of 70% suggests that the explicit discrepancy of forecasts with reality is not found:

$\Delta\sigma/\sigma$, %	*Forecast*	*Measurement* [91]
Cold-worked state	9–12	10
Recrystallized state	6–11	10

Predicted upon crystallographic texture the anisotropy of the strength to the plastic flow onset has the same appearance as the well-determined anisotropy of Young's technical elasticity modulus [84].

2. Simulation of mechanical tensile testing of a metal sheet. Simulating the deformation of test sample on a computer is the process of numerical solution of the equation of dynamics of a polycrystalline system with the cubic symmetry of crystals and rhombic texture at a given mechanical action.

The decision describes the evolution of the strain rate tensor of variously oriented crystals. Dynamic parameters of a system such as the tensors of plasticity of the hardening anisotropic crystals and the coefficients of stress relaxation in a polycrystalline environment with the changing shape of ellipsoidal grains, act as natural internal regulators of the deformation process.

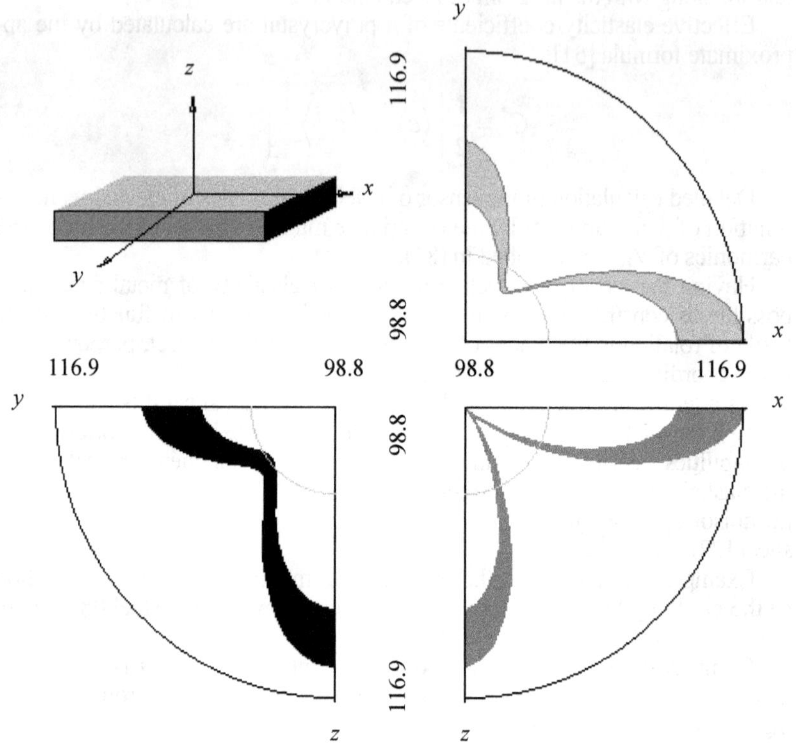

Fig. 5.3. Predicted anisotropy of the elasticity limit of cold-worked low-carbon steel (95% confidence interval) [MPa]

The observed tensor of the macroscopic strains rate is the integral variable of process depending on the probability distribution of the orientations of crystals. Evolution of the probability distribution of the orientations caused by rotation of the crystals lattice when constrained grain deformation in practice can be neglected (§ 4.1).

Information support of modeling the deformation of the metal sheet involves the distribution of the orientations of crystals, the physical properties of crystals (tensors of elasticity and plasticity) and the grain shape ratios $\left(a_x/a_z,\ a_y/a_z\right)$ in the sheet plane (*X*, *Y*) with the normal *Z*.

The finite-differences model of deformation processes in a polycrystalline system is designed for simulation tests on tension in the plane of metal sheet. Planned series of tests includes uniaxial tension in three directions relative to the rolling axis (0°, 45°, 90°) and symmetrical biaxial tension. Measure of deformation by uniaxial tension is an elongation of the sample, and on biaxial tensile a relative expansion of its area (or relative thinning).

A loading rate when simulating tensile tests is automatically specified but the researcher has the possibility to vary it in different series of tests provided it within acceptable limits for static tests controlled by the program.

At a fixed strain the dependence of all characteristics on the direction in a sheet plane is calculated by trigonometric interpolation over observation points. By the flowing components of the strain tensor it is determined the true coefficient of normal plastic anisotropy:

$$r(\varepsilon)=\left(\frac{db}{b}\right)\Big/\left(\frac{dh}{h}\right),$$

where b and h is width and thickness of the sample at a strain of ε.

Subsystem of studying the anisotropy of strength and plasticity of a metal sheet organizes simulating the mechanical tests, performs mathematical processing of results and creates the most complete graphical representation of the acquired information. Progress of mechanical tests is recorded to the protocol, which is stored in the archive.

3. Example of a comparative analysis of the anisotropy of thin metal sheets with the differing texture. As an instance, the specimens of steel 10GS and 08Yu are taken. Original data for analysis it is results of the optimum texture measurement performed by D.A. Kozlov (§ 2.3).

The form of identified texture of each tests specimen is visible over a random distribution of the orientations in a sampling aggregates of crystals that displayed in Fig. 5.4.

One of two reliably identified texture components in both specimens is of type $\{111\}\langle 110\rangle$ (in 08Yu its proportion about of 0.25, and in 10GS about of 0.9). The essential difference is that in 08Yu the crystallographic planes of $\{111\}$ are oriented well over sheet plane, whereas in 10GS there are ordered only crystallographic directions of $\langle 110\rangle$ along the rolling axis.

Simulating the tensile tests in the plane of sheet was carried out at a loading rate of 30 $\text{MPa}\cdot\text{s}^{-1}$. Physical and structural parameters on simulation of the deformation processes are the same as in Table 4.1. Information derived from experiments by the model is represented in Fig. 5.5 – 5.7

A picture of the anisotropy of the specimens studied by qualitatively is different. Strength to the plastic flow onset in the specimen of steel 08Yu proved to be larger in the rolling direction, while of steel 10GS transversely. Anisotropy of strain hardening has the same appearance as anisotropy of yield strength therefore with increasing deformation an overall picture does not change (Fig. 5.5).

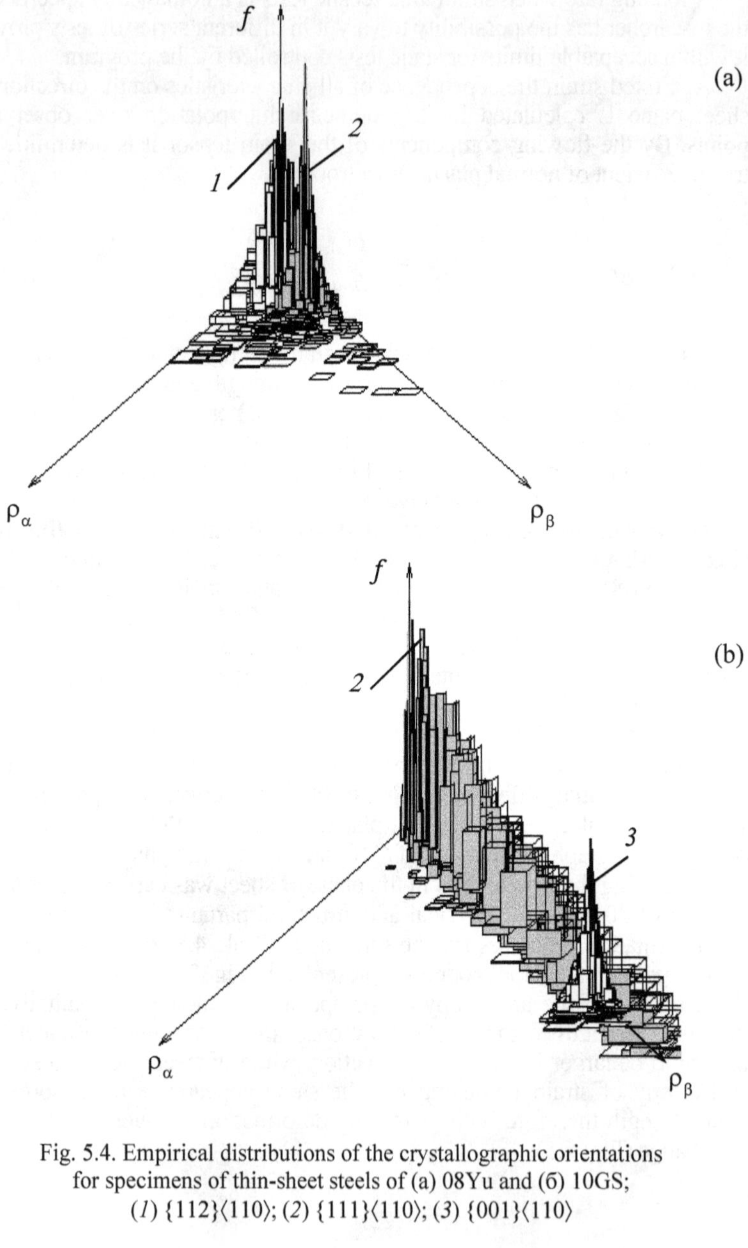

Fig. 5.4. Empirical distributions of the crystallographic orientations for specimens of thin-sheet steels of (a) 08Yu and (б) 10GS; (*1*) {112}⟨110⟩; (*2*) {111}⟨110⟩; (*3*) {001}⟨110⟩

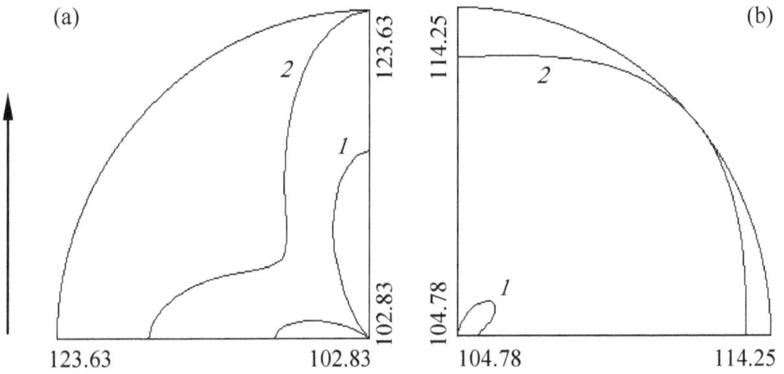

Fig. 5.5. Anisotropy of strength to plastic flow [MPa]: (a) 08Yu; (b) 10GS. The strain extent of (*1*) 0 and (*2*) 0.1%

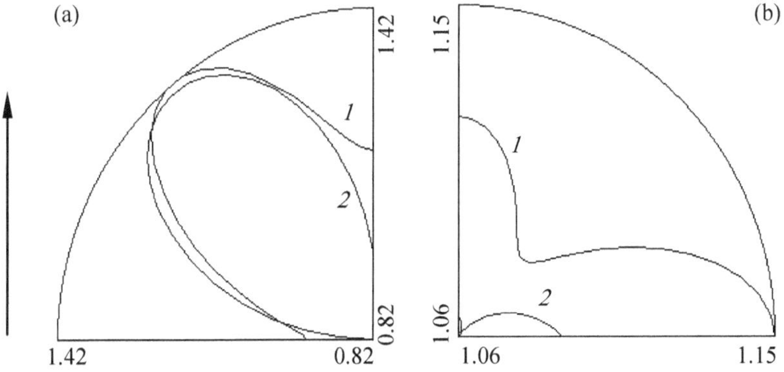

Fig. 5.6. Coefficient of normal plastic anisotropy: (a) 08Yu; (b) 10GS. The strain extent of (*1*) 0 and (*2*) 0.1%

Calculations show the rapid change of the strains anisotropy coefficient of $r(\varepsilon)$ yet before reaching the yield strength $\sigma_{0.1}$, which then is slow down.[1] Falling of value r_0 is greater for such direction in the sheet plane where the strength to the plastic flow onset of σ_0 is the greatest, that is, along the rolling in steel of 08Yu and across the rolling in steel of 10GS (Fig. 5.6).

[1] Similar results are available for copper from numerous measurements of the integral coefficient r_ε for different strains ε with subsequent extrapolation to $\varepsilon \to 0$ [96].

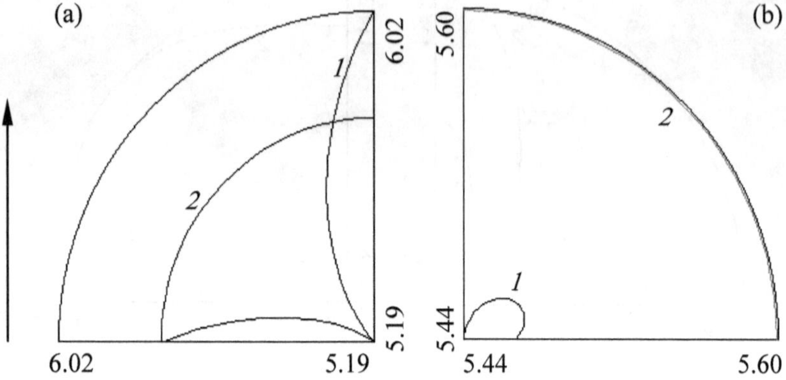

Fig. 5.7. Static viscosity [kJ/m³]: (a) 08Yu; (b) 10GS; (*1*) uniaxial tension; (*2*) symmetrical biaxial tension. The strain extent of 10%

When testing on biaxial tensile the properties of steel 08Yu take an intermediate position within the existing anisotropy. Otherwise behave the properties of steel 10GS, showing a particular sensitivity to the loading conditions. The difference is visible, for example, on the static viscosity that is measured by a specific work of deformation (Fig. 5.7).

Complete analysis of the information obtained during simulated tests of specimens of thin-sheet steels 08Yu and 10GS was carried out in [85].

4. Computer technology to studying the anisotropy of sheet materials. In-depth study of the anisotropy of the mechanical properties of metal sheet using the latest achievements of texture analysis contributes to the evidence-based assessment of its technological quality with the lowest cost for research.

Designed the automated system includes the tool means to ensure systematic accumulation of information and its transformation into a specific data requested by the user. Technology of research consists in building and using a knowledge base on the anisotropy of the properties of a broad class of sheet materials with cubic symmetry of crystal.

Software support of knowledge base provides:

- Organization in the disk file system of a computer the specified structures of data on the materials under study, and providing a list of specimens.
- Controlling the data updates developing a dynamic structure of data.
- Displaying the most complete achieved knowledge on the studied problem by the computer resources.

Adoption of the modern research technology to practice there puts forward a problem to automating the optimal by quality of object representation, control to measuring in the diffraction space.

PART II
DISLOCATION STRUCTURE OF STRONGLY DISTORTED CRYSTALS

In the outset there was a phenomenological model of a real structure of crystals with blocks and strains. By means of them the observed broadening of the X-ray lines was explained. Harmonic analysis of the profile of a few lines had to reveal a degree of imperfection of crystals [102].

With the creation of the theory of X-rays scattering, where used the dislocation model of structure of deformed crystals, it is made possible on width of the diffraction line to determine the average density of dislocations in deformed crystals. Applicable to practice formula exists for heavily washed-out nodes of the reciprocal lattice of crystals. Therefore, in principle, there is well measurable a very high density of randomly distributed direct dislocations when choosing a large length of the diffraction vector [39].

Achievements of the fundamental diffraction theory by M.A. Krivoglaz are summarized in [61]. There are mixed techniques, connecting parameters of the phenomenological model with the conditional density of dislocations, for example [1, 97].

Modern theory has advanced by the quality of the dislocation model of a real structure of crystals. Harmonic analysis of the diffraction line has renewed, reaching the optimal estimation of a random system of dislocations in crystals.

CHAPTER 6
THEORETICAL FOUNDATIONS OF DIFFRACTION STUDIES OF DISLOCATION STRUCTURE

System of dislocations occurring during the deformation of crystals is simulated by loops of a random size with a random distribution in its own slip planes. Diffraction theory constructed for the system of dislocation loops, is applicable to the entire space of structures, that includes disorder (random network), short-range order (random clusters), and long-range order (regular network). Observability of the dislocation structure depends on the existence of diffraction mappings that allow measuring the parameters of the model. Different states of a structure are measurable in the different areas of the diffraction space and with different accuracy [63, 72–73].

§ 6.1. Mathematical Description of the Diffraction by Deformed Crystals

Into the theory of diffraction are entered the Fourier representation of the displacement field from a circular dislocation loop, and the matrix of the spectral density of the distribution of loops in the crystal.

1. The diffraction equation for crystals with a large number of defects. Distribution of the intensity of X-ray scattering, when the atoms in a crystal are displaced from ideal lattice sites, has a Fourier image

$$I(\mathbf{R}) = F^2 \sum_{\mathbf{r}_s} \left\langle \exp\left[i\mathbf{q}\Delta\mathbf{U}(\mathbf{r}_s, \mathbf{R})\right]\right\rangle. \tag{6.1}$$

Here, F is the structural amplitude of scattering, $\mathbf{q}$ is the diffraction vector, $\Delta\mathbf{U}(\mathbf{r}_s, \mathbf{R})$ is the difference between the displacements of atoms from the $\mathbf{r}_s$ and $\mathbf{r}_s + \mathbf{R}$ sites at a given concentration and arrangement of defects, the summation is over all lattice sites, which number is N. Angle brackets denote averaging over the statistical ensemble, with all possible defects arrangement in a crystal at constant macroscopic parameters (density, correlation, order). It is considered that for such defects as dislocation, changes of the structural amplitude are negligible [39].

Displacement the lattice site $\mathbf{r}_s$ is combined out of displacements from all defects:

$$\mathbf{U}(\mathbf{r}_s) = \sum_{\mathbf{r}_t} c_\alpha(\mathbf{r}_t)\mathbf{u}_\alpha(\mathbf{r}_s - \mathbf{r}_t),$$

$c_\alpha(\mathbf{r}_t)$ is a random variable equal to unit if in site $\mathbf{r}_t$ there is a defect of type α, and zero if it there is not; $\mathbf{u}_\alpha(\mathbf{r}_s - \mathbf{r}_t)$ is displacement of atom out of lattice site $\mathbf{r}_s$ from the defect α, which is in site $\mathbf{r}_t$; for all types of defects α it is implied a summation.

Coefficients of the Fourier representation of the resulting displacement are of form

$$\mathbf{U}_{\mathbf{k}} = c_{\mathbf{k}}^{\alpha}\, \mathbf{u}_{\mathbf{k}}^{\alpha},$$

where $\mathbf{u}_{\mathbf{k}}^{\alpha} = \mathbf{u}_{-\mathbf{k}}^{*\alpha}$ and $c_{\mathbf{k}}^{\alpha} = c_{-\mathbf{k}}^{*\alpha}$ are the Fourier components of the displacement field $\mathbf{u}_{\alpha}(\mathbf{r})$ from defect of type α and the random function $c_{\alpha}(\mathbf{r})$, which describes the placement of these defects. What is more in $c_{\mathbf{k}}^{\alpha}$ there are represented the fluctuations in the number of defects by volume of crystal and local fluctuations in the statistical ensemble:

$$c_{\mathbf{k}}^{\alpha} = \left\langle c_{\mathbf{k}}^{\alpha} \right\rangle + \Delta c_{\mathbf{k}}^{\alpha}.$$

Equation (6.1) with the transition to the **k**- space takes the form

$$I(\mathbf{R}) = F^2 \sum_{\mathbf{r}_s} \left\langle \exp\left[i\mathbf{q} \frac{1}{N} \sum_{\mathbf{k}} c_{\mathbf{k}}^{\alpha}\, \mathbf{u}_{\mathbf{k}}^{\alpha} \left(1 - e^{i\mathbf{kR}}\right) e^{i\mathbf{kr}_s} \right] \right\rangle. \qquad (6.2)$$

Real crystal can be divided into elements of volume $\Delta V(\mathbf{r})$, each of which is much smaller the total volume V, but contains many defects. Under any detailed placement of defects in a statistical ensemble (at given macroscopic parameters) fluctuations of a local density in $\Delta V(\mathbf{r})$ will be small. Probability distribution of local fluctuations for a large number of defects in a crystal being approximated by the function [110]:

$$f = G \exp\left\{ -\frac{1}{2} \sum_{\mathbf{k}_1} \sum_{\mathbf{k}_2} \omega_{\mathbf{k}_1\mathbf{k}_2}^{\alpha\beta}\, \Delta c_{\mathbf{k}_1}^{\alpha}\, \Delta c_{\mathbf{k}_2}^{\beta} \right\},$$

$$\omega_{\mathbf{k}_1\mathbf{k}_2}^{\alpha\beta} = \left\langle \Delta c_{\mathbf{k}_1}^{\alpha} \Delta c_{\mathbf{k}_2}^{\beta} \right\rangle^{-1}$$

(G is the constant of normalization).

When period of fluctuations of the concentration field is much smaller than size of crystal, the matrix of the spectral density $\left\langle \Delta c_{\mathbf{k}_1}^{\alpha} \Delta c_{\mathbf{k}_2}^{\beta} \right\rangle$ is close to $\left\langle \Delta c_{\mathbf{k}}^{\alpha} \Delta c_{\mathbf{k}}^{\beta *} \right\rangle$ [30].

Averaging in the Eq. (6.2) by the probability of local fluctuations in the density of defects in a crystal results to the following diffraction equation:

$$\left.\begin{aligned}
I(\mathbf{R}) &= F^2 e^{-T} \sum_{\mathbf{r}_s} e^{i\Omega(\mathbf{r}_s)}, \\
T &= \frac{1}{N^2} \sum_{\mathbf{k}} \left(\mathbf{q}\mathbf{u}_{\mathbf{k}}^{\alpha}\right)\left\langle \Delta c_{\mathbf{k}}^{\alpha} \Delta c_{\mathbf{k}}^{\beta *}\right\rangle \left(\mathbf{q}\mathbf{u}_{\mathbf{k}}^{\beta}\right)^{*} (1-\cos \mathbf{kR}), \\
\Omega &= \frac{1}{N} \sum_{\mathbf{k}} \left(\mathbf{q}\mathbf{u}_{\mathbf{k}}^{\alpha}\right)\left\langle c_{\mathbf{k}}^{\alpha}\right\rangle \left(1-e^{i\mathbf{kR}}\right) e^{i\mathbf{k}\mathbf{r}_s}.
\end{aligned}\right\} \quad (6.3)$$

The matrix of the spectral density included in Eq. (6.3) is related to the macroscopic parameters of distribution of defects in a crystal.

2. Spectral density of defects distribution. Random values of $c_\alpha(\mathbf{r})$ are either zero or unity. Therefore

$$\left\langle c_\alpha(\mathbf{r}) c_\beta(\mathbf{r}+\boldsymbol{\rho})\right\rangle = \begin{cases} \left\langle c_\alpha(\mathbf{r})\right\rangle \delta_{\alpha\beta}, & \boldsymbol{\rho}=0, \\ \left\langle c_\alpha(\mathbf{r})\right\rangle P_{\beta|\alpha}(\mathbf{r},\boldsymbol{\rho}), & \boldsymbol{\rho}\neq 0. \end{cases}$$

Here, $\left\langle c_\alpha(\mathbf{r})\right\rangle$ and $P_{\beta|\alpha}(\mathbf{r},\boldsymbol{\rho})$ is the probability of occurrence defect α in the $\mathbf{r}$ and the conditional probability of occurrence defect β in the $\mathbf{r}+\boldsymbol{\rho}$ when in $\mathbf{r}$ there is of α ($\delta_{\alpha\beta}$ is Kronecker symbol).

When absence of order in arrangement of defects the probability of finding defect α in any lattice site $\mathbf{r}$ is the same: $\left\langle c_\alpha(\mathbf{r})\right\rangle = \overline{c_\alpha}$, where $\overline{c_\alpha} = \Pi_\alpha / N$; Π_α is number of defects α in a crystal. Introducing the correlation parameter $\varepsilon_{\alpha\beta}(\boldsymbol{\rho})$, characterizing average for all $\mathbf{r}$ probability of finding defects α and β at a distance of $\boldsymbol{\rho}$ ($\boldsymbol{\rho}\neq 0$, $\alpha=\beta$), it can be write

$$\left\langle c_\alpha(\mathbf{r}) c_\beta(\mathbf{r}+\boldsymbol{\rho})\right\rangle = \overline{c_\alpha c_\beta} + \varepsilon_{\alpha\beta}(\boldsymbol{\rho}),$$

$$\varepsilon_{\alpha\beta}(\boldsymbol{\rho}) = \overline{c_\alpha c_\beta}\,\frac{1-\overline{c}}{\overline{c}}\,\phi_{\alpha\beta}(\boldsymbol{\rho}),$$

and $\phi_{\alpha\beta}(\boldsymbol{\rho})$ is function of the dependence of correlation on the distance $\boldsymbol{\rho}$; $\overline{c}$ is total concentration of all defects (for $\overline{c} \ll 1$ negative correlation can be neglected, as its order of value $\overline{c}$) [44].

Description under the approximation of pair correlations is justified by the fact that defects in a crystal are found at microscopic distances from each other [101].

Let us suppose that among defects a certain proportion η_α forms a periodic structure with nodes $\mathbf{l}_m$, the total number of which is equal to M. Then

$$\langle c_\alpha(\mathbf{r})\rangle = \overline{c_\alpha}(1-\eta_\alpha) + \overline{c_\alpha}\eta_\alpha \frac{N}{M}\sum_{\mathbf{l}_m}\delta(\mathbf{r}-\mathbf{l}_m),$$

$$P_{\beta|\alpha}(\mathbf{r},\boldsymbol{\rho}) = \begin{cases} \dfrac{\overline{c_\beta\eta_\beta}}{\overline{c}}\dfrac{N}{M-1}\sum\limits_{\mathbf{l}_m}\delta(\boldsymbol{\rho}-\mathbf{l}_m),\ \mathbf{r}=\mathbf{l}_m, \\ c_\beta(\boldsymbol{\rho})(1-\eta_\beta) + c_\beta(\boldsymbol{\rho})\eta_\beta\dfrac{N}{M}\sum\limits_{\mathbf{l}_m}\delta(\mathbf{r}+\boldsymbol{\rho}-\mathbf{l}_m),\ \mathbf{r}\neq\mathbf{l}_m, \\ \qquad c_\beta(\boldsymbol{\rho}) = \overline{c_\beta} + \varepsilon_{\alpha\beta}(\boldsymbol{\rho})/\overline{c_\alpha}. \end{cases}$$

Here, $c_\beta(\boldsymbol{\rho})$ is concentration of β at a distance $\boldsymbol{\rho}$ from α; $\delta(\mathbf{r})$ is Dirac delta function.

Wave of density fluctuation of defects α over volume of crystal that considered being infinitely large,

$$\left.\begin{aligned} &\langle c_{\mathbf{k}}^{\alpha}\rangle = \sum_{\mathbf{r}}\langle c_\alpha(\mathbf{r})\rangle e^{i\mathbf{kr}} = \\ &= N\overline{c_\alpha}\left[(1-\eta_\alpha)\delta(\mathbf{k}) + \eta_\alpha\sum_{\mathbf{g}_m}\delta(\mathbf{k}-\mathbf{g}_m)\right], \end{aligned}\right\} \tag{6.4}$$

$\mathbf{g}_m$ is the reciprocal lattice vector of the ordered structure.

Using equations

$$\langle \Delta c_{\mathbf{k}}^{\alpha}\Delta c_{\mathbf{k}}^{\beta*}\rangle = \langle c_{\mathbf{k}}^{\alpha}c_{\mathbf{k}}^{\beta*}\rangle - \langle c_{\mathbf{k}}^{\alpha}\rangle\langle c_{\mathbf{k}}^{\beta}\rangle^*,$$

$$\langle c_{\mathbf{k}}^{\alpha}c_{\mathbf{k}}^{\beta*}\rangle = N\overline{c_\alpha}\delta_{\alpha\beta} + \sum_{\mathbf{r}}\sum_{\boldsymbol{\rho}\neq 0,\ \alpha=\beta}\langle c_\alpha(\mathbf{r})c_\beta(\mathbf{r}+\boldsymbol{\rho})\rangle\, e^{i\mathbf{k}\boldsymbol{\rho}}$$

it is possible to determine the spectral density of distribution of defects:

$$\left.\begin{aligned} &\langle \Delta c_{\mathbf{k}}^{\alpha}\Delta c_{\mathbf{k}}^{\beta*}\rangle = \\ &= N\left\{\overline{c_\alpha} - \overline{c_\alpha c_\beta}(1-\eta_\alpha)(1-\eta_\beta)\right\}\delta_{\alpha\beta} + \\ &+ N\overline{c_\alpha c_\beta}(1-\eta_\alpha\eta_\beta)\frac{1-\overline{c}}{\overline{c}}\sum_{\boldsymbol{\rho}\neq 0,\ \alpha=\beta}\phi_{\alpha\beta}(\boldsymbol{\rho})e^{i\mathbf{k}\boldsymbol{\rho}} + \\ &+ N^2\overline{c_\alpha c_\beta}\,\eta_\alpha\eta_\beta\frac{1-\overline{c}}{\overline{c}}\sum_{\mathbf{g}_m=0,\ \alpha\neq\beta}\delta(\mathbf{k}-\mathbf{g}_m). \end{aligned}\right\} \tag{6.5}$$

According to the spectral density expression there are the following components of T in the diffraction equation of (6.3):

$$\left.\begin{aligned} T_0 &= q_i q_j \left[\overline{c_\alpha} - \overline{c_\alpha c_\beta}\left(1-\eta_\alpha\right)\left(1-\eta_\beta\right)\right] \times \\ &\times \left[\frac{1}{N}\sum_{\mathbf{k}} W_{ij}^{\alpha\beta}\left(\mathbf{k},\mathbf{R}\right)\right] \delta_{\alpha\beta}, \end{aligned}\right\} \quad (6.6)$$

$$\left.\begin{aligned} T_1 &= q_i q_j \overline{c_\alpha c_\beta}\left(1-\eta_\alpha \eta_\beta\right) \times \\ &\times \frac{1-\overline{c}}{\overline{c}}\left[\frac{1}{N}\sum_{\mathbf{k}} W_{ij}^{\alpha\beta}\left(\mathbf{k},\mathbf{R}\right) \sum_{\rho\neq 0,\ \alpha=\beta} \phi_{\alpha\beta}\left(\boldsymbol{\rho}\right) e^{i\mathbf{k}\boldsymbol{\rho}}\right], \end{aligned}\right\} \quad (6.7)$$

$$T_2 = q_i q_j \overline{c_\alpha c_\beta}\, \eta_\alpha \eta_\beta \frac{1-\overline{c}}{\overline{c}}\left[\sum_{\mathbf{g}_m} W_{ij}^{\alpha\beta}\left(\mathbf{g}_m,\mathbf{R}\right)\right] \delta_{\alpha\beta}, \quad (6.8)$$

$$W_{ij}^{\alpha\beta}\left(\mathbf{k},\mathbf{R}\right) = u_i^\alpha\left(\mathbf{k}\right) u_j^\beta\left(\mathbf{k}\right)^* \left(1-\cos\mathbf{kR}\right),$$

where T_0 corresponds to a random distribution of defects, T_1 takes into account a correlation, and T_2 means the appearance of periodicity.

When assumed a random distribution of defects in crystal, so that at any point $\mathbf{r}$

$$c_\alpha\left(\mathbf{r}\right) = \begin{cases} 1 \text{ with probability } \overline{c_\alpha}, \\ 0 \text{ with probability } 1-\overline{c_\alpha}, \end{cases}$$

and moreover $\overline{c} \ll 1$, the original equation for $I(\mathbf{R})$ yields the formula obtained in [39]:

$$T = \overline{c}\sum_{\mathbf{r}_s}\left\{1-\exp\left[i\mathbf{q}\Delta\mathbf{u}\left(\mathbf{r}_s,\mathbf{R}\right)\right]\right\}.$$

If to introduce the Fourier representation for $\Delta\mathbf{u}\left(\mathbf{r},\mathbf{R}\right)$ and to assume that $\mathbf{q}\,\Delta\mathbf{u}_{\mathbf{k}}\left(\mathbf{R}\right)/N \ll 1$, then after summation over $\mathbf{r}_s$ it is obtained

$$T = \frac{\overline{c}}{2}\left[\frac{1}{N}\sum_{\mathbf{k}}\left|\mathbf{q}\,\Delta\mathbf{u}_{\mathbf{k}}\left(\mathbf{R}\right)\right|^2\right],$$

that coincides with $T = T_0$, where $\eta_\alpha = \eta_\beta = 0$, $\varepsilon_{\alpha\beta} = 0$, and all $\overline{c_\alpha} \ll 1$.

Waves of density fluctuations of defects in a crystal of $\left\langle c_{\mathbf{k}}^{\alpha}\right\rangle$ affect the function Ω in the Eq. (6.3). Assuming that $\left\langle c_{\mathbf{k}}^{\alpha}\right\rangle$ is of the form (6.4), we have

$$\sum_{\mathbf{r}_s} e^{i\Omega(\mathbf{r}_s)} = N\left[1-\overline{c_\alpha}\eta_\alpha\left(1-e^{i\mathrm{H}}\right)\right],$$

$$\mathrm{H}=\overline{c_\alpha}\eta_\alpha\sum_{\mathbf{g}_m}\left(\mathbf{q}\mathbf{u}_{\mathbf{g}_m}^{\alpha}\right)\left(1-e^{i\mathbf{g}_m\mathbf{R}}\right).$$

So, the sine harmonics of the diffraction line that appear with the defects ordering are negligible, their amplitude is of $\sim \bar{c}$ from the cosine harmonics.

Constructed the diffraction equation applies subject to small compared to a crystal size periods of fluctuations as the concentration field of defects, and created by them the field of lattice displacements.

3. The displacements field in a crystal with dislocation loops. Type of dislocation loops is completely determined by the tensor of dislocation moment $d_{ij} = b_i S_j$, where b_i is i-th component of the Burgers vector, S_j is the projection area of the loop on a plane perpendicular to the j-th coordinate axis.

In continuum approximation vector of displacement at a point $\mathbf{r}$ from a closed dislocation loop with center $\mathbf{r}'$ is

$$u_m(\mathbf{r}) = b_i\int_S \lambda_{ijkl}\frac{\partial}{\partial x_l}\hat{G}_{km}(\mathbf{r}-\mathbf{r}')\,dS_j',$$

λ_{ijkl} is tensor of elastic moduli, and $\hat{G}_{km}(\mathbf{r})$ is Green's tensor-function; the integration runs over the loop area S [45].

From here the Fourier component of the displacements field of circular dislocation loop will be

$$u_m(\mathbf{k}) = -2\frac{i}{v_0}\lambda_{ijkl}\,d_{ij}\,k_l\,G_{km}(\mathbf{k})\frac{J_1(k'\xi)}{k'\xi},$$

where $G_{km}(\mathbf{k})$ is the Fourier-image of Green function, $J_n(x)$ is cylindrical Bessel functions of I kind, k' is modulus of projection of the wave vector $\mathbf{k}$ on loop plane, ξ is loop radius, and v_0 is volume of the unit cell of crystal.

If to employ obtained in the approximation of elastic isotropy of a crystal the expression of $G_{km}(\mathbf{k})$ [18], then

$$u_m(\mathbf{k}) = -2\frac{i}{v_0}\left(\frac{k_l}{k^2}D_{lm}-\sigma_1\frac{k_k k_l k_m}{k^4}D_{kl}\right)\frac{J_1(k'\xi)}{k'\xi},$$

$$D_{ij} = \sigma_2 d_{ll}\delta_{ij} + \left(d_{ij} + d_{ji}\right) \ \left(\sigma_1 = 1/2(1-\sigma),\ \sigma_2 = 2\sigma/(1-2\sigma)\right),$$

σ is Poisson coefficient, and δ_{ij} is Kronecker symbol.

Since for dislocation loops $|\mathbf{u}_\mathbf{k}| \sim \frac{|\mathbf{k}|}{k^2}\frac{1}{v_0}bS$, then under $|\mathbf{k}| \geq \frac{2\pi}{\Xi}$ and $|\mathbf{q}| = \frac{2\pi}{a}\sqrt{H^2+K^2+L^2}$, where a is period of the crystal lattice, Ξ is crystal size, and $\{HKL\}$ is reflection indexes, there is restrictions

$$\frac{1}{N}\mathbf{q}\,\Delta\mathbf{u}_\mathbf{k}(\mathbf{R}) < \frac{1}{V}\frac{\Xi}{a}bS\sqrt{H^2+K^2+L^2} < \left(\frac{2\xi}{\Xi}\right)^2\sqrt{H^2+K^2+L^2}.$$

Consequently, the diffraction line harmonics will be calculated with good accuracy when the size of dislocation loops on the order of less than the crystal size.

4. A Bounding surface to crystal. The displacements field in a crystal of finite size should not create stresses on bounding its volume V surface S_V. Condition is satisfied if

$$U_j(\mathbf{r}) = U_j^\infty(\mathbf{r}) - \overline{U_j^\infty}(\mathbf{r}) + e_{ij}r_i,$$

where U_j^∞ are displacements in an infinite elastic medium, when inside allocated surface S_V are introduced defects; $-\overline{U_j^\infty}$ are displacements that return the size and shape of S_V to the original; e_{ij} is tensor of uniform extension when removing the compression by non-deformed matrix [18].

Equation (6.1) for the finite-size crystal takes the form

$$I(\mathbf{R}) = F^2 e^{iq_j e_{ij} R_j} \times$$
$$\times \sum_{\mathbf{r}_s} \exp\left[-iq_j\Delta\overline{U_j^\infty}(\mathbf{r}_s,\mathbf{R})\right]\left\langle \exp\left[iq_j\Delta U_j^\infty(\mathbf{r}_s,\mathbf{R})\right]\right\rangle.$$

Last factor was calculated and presented in Eq. (6.3). Uniform lattice extension of e_{ij} leads to shift of lines without changing their shape. It remains to re-determine the sum over $\mathbf{r}_s$ that adjusted taking into account the $-\overline{U_j^\infty}$.

For crystal without macroscopic bending (sum of Burgers vectors equal to zero), the displacements $-\overline{U_j^\infty}(\mathbf{r})$ are determined by using the equation of elastic equilibrium in the presence of solid forces that are expressed through tensor of the dislocation moment density [45]:

$$P_{ik} = \frac{j_\alpha d_{ik}^\alpha}{V}\,\delta(V).$$

Here, j_α is number of loops with tensor of the dislocation moment d_{ik}^α in area V bounded by surface S_V (over all α implied summation). Besides in an infinite space is introduced a crystal shape function:

$$s(\mathbf{r}) = v_0 \sum_{\mathbf{r}_t} \delta(\mathbf{r} - \mathbf{r}_t).$$

Solution of the problem in the Fourier representation, where $s(\mathbf{r})$ has the components $s_\mathbf{k}$, will be

$$\overline{U_j^\infty}(\mathbf{k}) = -\overline{c}_\alpha \frac{i}{v_0} \lambda_{iklm} d_{ik}^\alpha k_m G_{lj}(\mathbf{k})\, s_\mathbf{k}.$$

For a bounded crystal is modifying the wave of fluctuation of defects density from Eq. (6.4): $\langle c_\mathbf{k}^\alpha \rangle = \overline{c}_\alpha s_\mathbf{k}$ $(\eta_\alpha = 0)$. Subject to the necessary corrections there are received formulas

$$\Omega(\mathbf{r}_s) = q_j \overline{c}_\alpha \frac{1}{N} \sum_\mathbf{k} h_j^\alpha(\mathbf{k}) \left(1 - e^{i\mathbf{k}\mathbf{R}}\right) s_\mathbf{k}\, e^{i\mathbf{k}\mathbf{r}_s},$$

$$h_j^\alpha(\mathbf{k}) = -\frac{i}{v_0}\left(\frac{k_m}{k^2} D_{mj} - \sigma_1 \frac{k_l k_m k_j}{k^4} D_{lm}\right)\left(2\frac{J_1(k'\xi)}{k'\xi} - 1\right). \qquad (6.9)$$

If the Fourier components $|\Omega_\mathbf{k}| \ll 1$, then

$$\left.\begin{aligned} &\sum_{\mathbf{r}_s} e^{i\Omega((\mathbf{r}_s)} \cong N\left[e^{-\omega} + i\beta\right], \\ &\omega = \frac{1}{N^2} \sum_\mathbf{k} \left|q_j \overline{c}_\alpha h_j^\alpha(\mathbf{k})\right|^2 (1 - \cos \mathbf{k}\mathbf{R})\, |s_\mathbf{k}|^2, \\ &\beta = q_j \overline{c}_\alpha \frac{1}{N^2} \sum_\mathbf{k} h_j^\alpha(\mathbf{k}) \left(1 - e^{i\mathbf{k}\mathbf{R}}\right) |s_\mathbf{k}|^2. \end{aligned}\right\} \qquad (6.10)$$

For interval $0 < |k| < 2\pi/\Xi$ is obtained an estimate $|\Omega_\mathbf{k}| \ll \overline{c}(\xi/a)^3$.

Thus, the formula is acceptable, as long as the dislocation density $\rho_d \leq 2\pi/\xi^2$ (for example, $\rho_d \leq 10^{11}$ cm^{-2} when $\xi = 500\,a$).

The consequence of finite sizes of crystal it is attenuation of cosine harmonics in the $I(\mathbf{R})$ and appearance of sine harmonics. As a result, the distribution of scattering intensity is made even more diffuse and asymmetrical.

In Equation (6.3) quantity ω gives an increase in T which relative value is of the order of $\bar{c}$. Therefore with sufficient accuracy the form of intensity distribution is represented in the Fourier components

$$I(\mathbf{R}) \cong NF^2 e^{-T}\left[1 + i\beta\right],$$

where β is the relative value of sine harmonics from the Eq. (6.10).

5. A Random system of dislocations in the real structure of crystals. It is reasonably to suggest that in deformed crystals in each of p existing slip systems arises a random number of dislocation loops Π_α ($\alpha = 1, \ldots, p$) with random sizes ξ and coordinates $\mathbf{r}$. Type of loops, that is dislocation Burgers vector $\mathbf{b}$ and vector normal to the loop plane $\mathbf{n}$, is determined by slip system so that Π_α there is a random number of loops of type α.

Let us assume that all the random variables describing the statistical ensemble of dislocations systems are mutually independent. It should be continue to average the e^{-T} over the ensemble that has received additional degrees of freedom. At first we will find the average $\left\langle e^{-T}\right\rangle$ over all possible Π_α with the total number of loops in the crystal of $\Pi = \sum_{\alpha=1}^{p}\Pi_\alpha$. Further averaging over random realizations of $\xi(\mathbf{r})$ in the ensemble will be replaced on averaging over the distribution of ξ in the volume of the crystal, so as is allowed by the ergodic hypothesis [44].

With a large number of loops Π the fluctuations in their local density and the sizes distribution with respect to the averages over the ensemble will be small. Then fluctuations of the quantity T as well will be small, so we can write $\left\langle e^{-T}\right\rangle \cong e^{-\langle T\rangle}$ [37].

Let the dislocation loops with equal probability can be located in all p slip systems in a crystal, which is close to reality for bcc crystals and feasibly in the fcc crystals with the high energy of stacking fault. The probability that exactly Π_α ($\alpha = 1, \ldots, p$) loops from the total number of Π, which has approximately the Poisson distribution, are found in the slip system α, is determined by Bernoulli distribution $B_{1/p}\left(\Pi, \Pi_\alpha\right)$ [105]. Then the expected concentration of loops in each slip system is equal to $\bar{c}/p$, where $\bar{c} = \Pi/N$ is the mean concentration of all dislocation loops in a crystal.

When calculating T the sum over $\mathbf{k}$ in Eq. (6.6) and (6.7) can be substituted by an integral over $\mathbf{k}$-space multiplied by a normalizing constant $V/(2\pi)^3$. Having in mind that $\bar{c} \ll 1$, we obtain the equation

$$\left.\begin{aligned} T_0 &= q_i q_j \bar{c}\left[\frac{1}{p}\sum_{\alpha=1}^{p}\Gamma_{ij}^{(\alpha\beta)}\left\langle(\xi/a)^4\,\Phi_{\alpha\beta}(\mathbf{R},\xi)\right\rangle_{\xi}\delta_{\alpha\beta}\right], \\ \Phi_{\alpha\beta} &= \left(a/2\pi\right)^3\int_{\mathbf{k}} w_{\alpha\beta}(\mathbf{k})\,d\mathbf{k}; \end{aligned}\right\} \quad (6.11)$$

$$w_{\alpha\beta}(\mathbf{k}) = \left(\frac{J_1(k_\alpha{}'\xi)}{k_\alpha{}'\xi}\frac{J_1(k_\beta{}'\xi)}{k_\beta{}'\xi}\right)\frac{1}{k^2}\left(1-\cos\mathbf{kR}\right),$$

$$\begin{cases} \Gamma_{ij}^{(\alpha\beta)} = \mathrm{B}_{klmn}^{(\alpha\beta)}E_{ijklmn}, \\ \mathrm{B}_{klmn}^{(\alpha\beta)} = (2\pi/a)^2\left(b_k^\alpha n_l^\alpha + b_l^\alpha n_k^\alpha\right)\left(b_m^\beta n_n^\beta + b_n^\beta n_m^\beta\right), \\ E_{ijklmn} = \delta_{ik}\delta_{jm}\delta_{ln} - \sigma_1(\delta_{ik}e_{jmln} + \delta_{jm}e_{ikln}) + \sigma_1^2 e_{ijklmn}. \end{cases}$$

Symbols e_{ijkl} and e_{ijklmn} are equal to unit, when the number of identical indexes is even, and zero when it is odd.

At the tensor coefficients $\Gamma_{ij}^{(\alpha\beta)}$ is represented the destroying power of the dislocation field for periodic structure of crystals: the higher the coefficients, the more diffuse nodes of the reciprocal lattice upon any shape.

Averaging over the distribution of loop sizes hereinafter is denoted by angle brackets marked with symbol ξ.

Number of loops in the slip systems (α, β) under their total number in crystal of Π is described by correlated random values obeying the bivariate Bernoulli distribution [105]. On the assumption that sizes of all loops, whose number within limitations $1 \ll \Pi \ll N$, are independent identically distributed random values we obtain the equation of the averaged correlation component in the form

$$\left.\begin{aligned} T_1 &= q_i q_j \bar{c}\left(1-\bar{\eta}\right)\times \\ &\times\left[\frac{1}{p^2}\sum_{\alpha=1}^{p}\sum_{\beta=1}^{p}\Gamma_{ij}^{(\alpha\beta)}\left\langle(\xi/a)^4\,\Psi_{\alpha\beta}(\mathbf{R},\xi)\right\rangle_{\xi}\right], \\ \Psi_{\alpha\beta} &= \left(a/2\pi\right)^3\int_{\mathbf{k}} w_{\alpha\beta}(\mathbf{k})\,\phi_{\mathbf{k}}^{\alpha\beta}\,d\mathbf{k}; \end{aligned}\right\} \quad (6.12)$$

$\phi_{\mathbf{k}}^{\alpha\beta}$ is Fourier image of the correlation function $\phi_{\alpha\beta}(\boldsymbol{\rho})$, depending on the distance $\boldsymbol{\rho}$ between the centers of loops; $\underline{\eta}$ is proportion of all the orderly distributed loops from the total number of Π.

When calculating the periodic component of the T_2, defined by Eq. (6.8), summation over all space of the reciprocal lattice ordered structure is substituted by summation over unit cell vectors of $\mathbf{g} = \left(\frac{2\pi}{\ell}\right)\mathbf{j}$, where ℓ is period of main translations, $\mathbf{j}$ are unit vectors in the translation directions. To take into account the number of cells of the reciprocal lattice of ordered defects in a unit volume of the reciprocal space of crystal, each term is multiplied by $\left(\frac{a}{\ell}\right)^3$.

The equation of the averaged periodic component will have the following form

$$\left.\begin{aligned} T_2 &= \bar{q}_i \bar{q}_j \bar{c}\eta \left[\frac{1}{p^2}\sum_{\alpha=1}^{p}\sum_{\beta=1}^{p}\left\langle(\xi/a)^4\,\Theta_{\alpha\beta}(\mathbf{g},\mathbf{R})\right\rangle_{\xi}\right], \\ \Theta_{\alpha\beta} &= \sum_{\mathbf{j}}\gamma_{ij}^{(\alpha\beta)}\left\langle\left(\frac{a}{\ell}\right)^3 w_{\alpha\beta}\left(\frac{2\pi}{\ell}\mathbf{j}\right)\delta_{\alpha\beta}\right\rangle_{1/\ell}, \end{aligned}\right\} \tag{6.13}$$

$$\gamma_{ij}^{(\alpha\beta)} = \left(t_i^{(\alpha)}t_j^{(\beta)}\right),\; t_i^{(\alpha)} = (2\pi/a)\left(b_k^{\alpha}n_l^{\alpha} + b_l^{\alpha}n_k^{\alpha}\right)\left(\delta_{ik}j_l - \sigma_1 j_i j_k j_l\right).$$

Constructed the diffraction equation conforms to the structure of deformed crystals that modeled by the distribution of dislocation loops in their own slip planes.

By available estimates of dislocation density 10^8 cm^{-2} < $\rho_d < 10^{13}$ cm^{-2}, where $\rho_d = \left(2\pi/a^2\right)\left[\bar{c}\left(\bar{\xi}/a\right)\right]$ the average concentrations of dislocation loops should be in the range $10^{-8}\left(\bar{\xi}/a\right)^{-1} < \bar{c} < 10^{-3}\left(\bar{\xi}/a\right)^{-1}$. Thus the smallest of possible number of loops in crystals of the smallest existing size ($\Xi \sim 10$ μm) has a lower limit $\Pi > 10^3$.

The theory assumptions, which are justified for a large number of defects, are applicable to the object under study.

§ 6.2. Predicted Diffraction Pattern for System of the Dislocation Loops

When studying the data predicted by the equation of diffraction there is found out the role of various parameters of dislocations system upon real structure of crystals.

1. Harmonics of the diffraction line. To calculate harmonics of the $\{HKL\}$ diffraction line Z axis of the Cartesian coordinates system (X, Y, Z) is chosen along the normal to the reflecting plane $\{HKL\}$. In this coordinate basis

$$\mathbf{q} = \left[0, 0, \frac{2\pi}{a}\sqrt{H^2 + K^2 + L^2}\right], \quad \mathbf{R} = \left[0, 0, m\frac{a}{\sqrt{H^2 + K^2 + L^2}}\right].$$

When a crystal rotates around the Z axis, the effect of atomic displacements in the reflecting plane itself disappears in sum, and the distribution of scattering intensity along $\mathbf{q}$ depends only on displacements normal to the $\{HKL\}$ plane.

Of Equations (6.11) – (6.13) at given $\{HKL\}$, $T(m)$ $(m = 1, 2, \ldots)$ is calculated as the minus logarithm of the normalized harmonics $e^{-T(m)}$ of order m.

As example the $\{112\}$ diffraction line of bcc iron crystals is selected. Only the slip set $\langle 111\rangle\{110\}$ is believed to be active. Specified the dislocation density is 10^{10} cm^{-2}.

2. Effect of sizes level of the dislocation loops. Consider the model of a random distribution of dislocation loops, when $T = T_0$. In Eq. (6.11) of component T_0 are summed up $\Phi_{\alpha\beta}$ for $\alpha = \beta$. To integrate in **k**-space it is suitable a cylindrical system of coordinate associated with the normal vector **n** to the plane of loop α:

$$\int_{\mathbf{k}} w_{\alpha\beta}(\mathbf{k})\, d\mathbf{k} = 2\pi^2 Z \delta_{\alpha\beta},$$

$$\left.\begin{aligned} Z &= \int_0^{\infty} \left[J_1(\xi t)/(\xi t)\right]^2 \left[1 - e^{-ut} J_0(vt)\right] dt, \\ u &= (\mathbf{R}\cdot\mathbf{n}), \quad v = \sqrt{R^2 - (\mathbf{R}\cdot\mathbf{n})^2}. \end{aligned}\right\} \qquad (6.14)$$

Computed $T(m) = T_0(m)$ at the same size of dislocation loops ξ, equated to the average $\bar{\xi}$, is shown in Fig. 6.1.

Type curve $T_0(m)$ depends on the size level of the dislocation loops: $T_0 \sim (1 - e^{-m})$ when the small loops ($\xi \le 50\, a$); $T_0 \sim m^{\gamma}$ $(\gamma < 1)$ for medium-sized loops ($50\, a < \xi \le 200\, a$); $T_0 \sim m$ with the large loops ($\xi > 200\, a$).

General picture of the effect of the dislocation loops sizes on the harmonics of the diffraction line is the same as in the theory by M.A. Krivoglaz [39].

Specific numerical values are close until loops yet small [63]. Formula for displacements of the lattice sites at distances r from the dislocation loop, which is used in [39], it is $|\mathbf{u}(\mathbf{r})| \sim b\xi^2/r^2$ $(r \gg \xi)$ [18]. An increase in the radius of loops ξ, when the same dislocation density in crystal reduces the number of sites sufficiently far from all loops for acceptable accuracy of calculation.

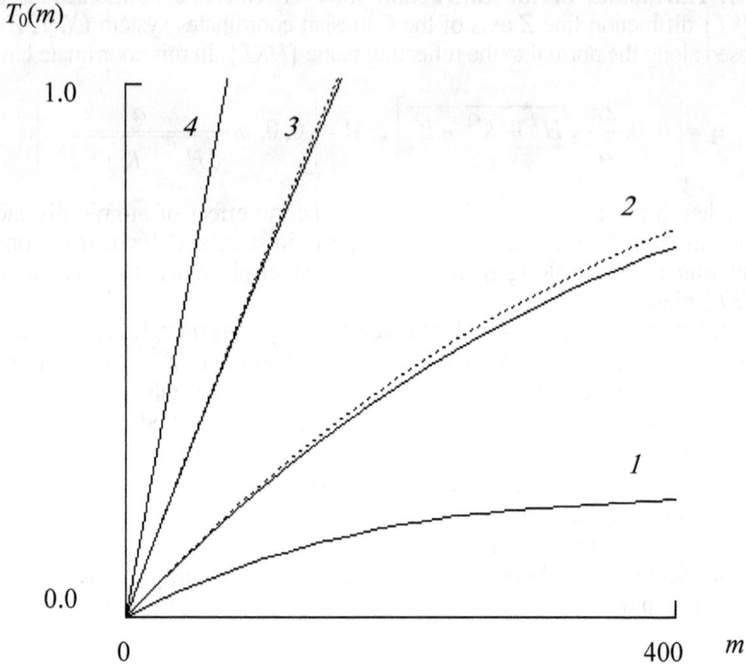

Fig. 6.1. Curves of the logarithms of normalized harmonics of the diffraction line on changing mean sizes of dislocation loops: (*1*) $\xi = 50\ a$; (*2*) $\xi = 100\ a$; (*3*) $\xi = 250\ a$; (*4*) $\xi = 500\ a$. Dashed lines with allowance for the dispersion of the loops sizes about the mean

In order to approximately allow for inhomogeneity of the dislocation loops sizes, we assume the logarithmically normal distribution of ξ. (It is predictably asymmetric distribution of the probabilities of sizes of loops, since their expansion encounters obstacles.) With logarithmically normal distribution a random value ξ can vary by more than an order of magnitude if the coefficient of variation $\sigma_\xi / \bar{\xi} > 1/3$.

Curves of the function $T_0(m)$ with allowance for the dispersion of ξ relative to $\bar{\xi}$ are plotted by the dashed line in Fig. 6.1. The given coefficient of variation of the loops sizes it is $\sigma_\xi / \bar{\xi} = 0.2$. Averaging over random fluctuations of ξ was performed by the Monte Carlo method. Deviations of the dashed lines suggest that the inhomogeneity of the loops sizes speed up the decrease of harmonics of the diffraction line.

3. Models of correlation in the dislocation structure of deformed crystals. According to the physical representations about the object, a random pile-up of dislocation loops arises because of correlation in the distribution of loops in the slip plane itself and along parallel planes. (Correlation on the intersection line of the slip planes is neglected as its contribution is small, and the calculation is too complicated). To model a pile-up, it is required to construct a convolution of two correlation functions, one of which depends on the distance in the plane of the loop, the other from the distance along the normal to it.

Let us denote the functions of normal (one-dimensional) and planar (two-dimensional) correlation as $\phi^{(1)}(\boldsymbol{\rho})$ and $\phi^{(2)}(\boldsymbol{\rho})$. The space (three-dimensional) correlation function and its Fourier image are determined by the formulas

$$\phi^{(3)}(\rho) = \int_{-\infty}^{\infty} \phi^{(1)}(\zeta)\, \phi^{(2)}(\rho - \zeta)\, d\zeta, \quad \phi_{\mathbf{k}}^{(3)} = \phi_{\mathbf{k}}^{(1)}\, \phi_{\mathbf{k}}^{(2)}.$$

It can be supposed that the correlation decreases with distance $\boldsymbol{\rho}$ between the centers of the loops in accordance with the exponential law.[1] Arguments of the exponent are in $\phi^{(1)}(\boldsymbol{\rho})$ the projection of $\boldsymbol{\rho}$ on normal vector $\mathbf{n}$ to the loop plane, and in $\phi^{(2)}(\boldsymbol{\rho})$ the length of projection vector of $\boldsymbol{\rho}$ on the plane:

$$\phi^{(1)}(\boldsymbol{\rho}) = \exp\left\{-\frac{\boldsymbol{\rho}\cdot\mathbf{n}}{\mu}\right\}\, \delta(\mathbf{n}\times[\boldsymbol{\rho}\times\mathbf{n}]),$$

$$\phi^{(2)}(\boldsymbol{\rho}) = \exp\left\{-\frac{|\mathbf{n}\times[\boldsymbol{\rho}\times\mathbf{n}]|}{\tau}\right\}\, \delta(\boldsymbol{\rho}\cdot\mathbf{n}).$$

Parameters μ and τ denote distances within which the probability of finding a pair of loops is appreciably higher than with their chaotic distribution over the volume of the crystal.

In Equation (6.12) of T_1 component for chosen correlation function following expression will appear

$$\int_{\mathbf{k}} w_{\alpha\beta}(\mathbf{k})\, \phi_{\mathbf{k}}^{\alpha\beta}\, d\mathbf{k} = 2\pi^2 X\, \delta_{\alpha\beta},$$

---

[1] If β arises at a distance $\boldsymbol{\rho}_1$ from α with a probability $\phi(\boldsymbol{\rho}_1)$, and γ arises at a distance $\boldsymbol{\rho}_2$ from β with a probability $\phi(\boldsymbol{\rho}_2)$, then as a result of two independent random events γ arises at the distance $\boldsymbol{\rho} = \boldsymbol{\rho}_1 + \boldsymbol{\rho}_2$ from α with the probability $\phi(\boldsymbol{\rho}) = \phi(\boldsymbol{\rho}_1)\,\phi(\boldsymbol{\rho}_2)$. The function that satisfies this equation it is $\phi(\boldsymbol{\rho}) \sim e^{h\boldsymbol{\rho}}$ (h is the bonding parameter).

$$\left.\begin{aligned}
&X = 2\left[(\mu/d)(\tau/b)^2\right]E,\\
&E = \int_0^\infty \left(\frac{J_1(\xi t)}{\xi t}\right)^2 \frac{F_1(t)-\mu t F_2(t)}{\left[1-(\mu t)^2\right]\left[1+(\tau t)^2\right]^{3/2}}\,dt,\\
&F_1(t) = 1-e^{-ut}J_0(vt),\; F_2(t) = 1-e^{-u/\mu}J_0(vt);
\end{aligned}\right\} \quad (6.15)$$

d and b are the interplanar and interatomic distances (u and v are the same variables as in Eq. (6.14)).

Correlation component T_1 when different parameters of the model of the dislocation structure of iron crystals in the absence of long-range order $(\bar{\eta} = 0)$ is shown in Fig. 6.2.

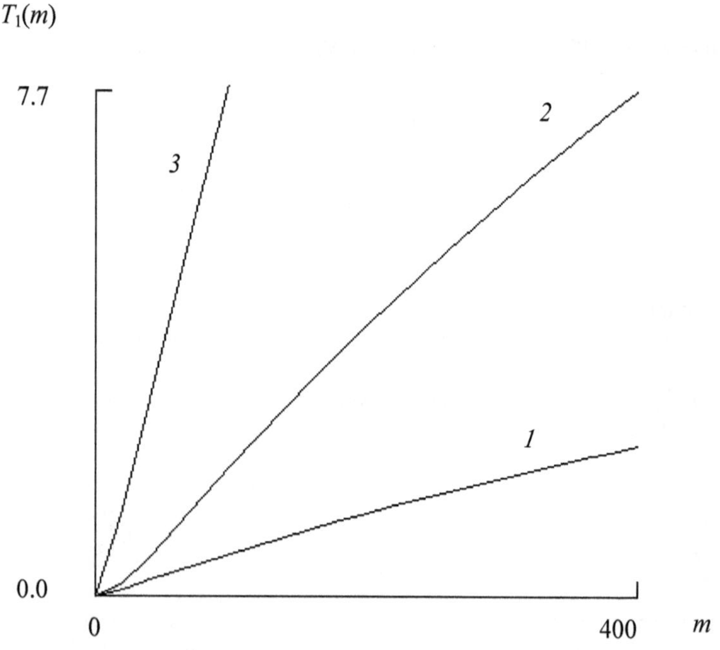

Fig. 6.2. Correlation component of the harmonics logarithm depending on the model parameters: (*1*) $\tau/b = 2$, $\mu/d = 2$; (*2*) $\tau/b = 2$, $\mu/d = 7$; (*3*) $\tau/b = 7$, $\mu/d = 2$; $\xi = 100\ a$

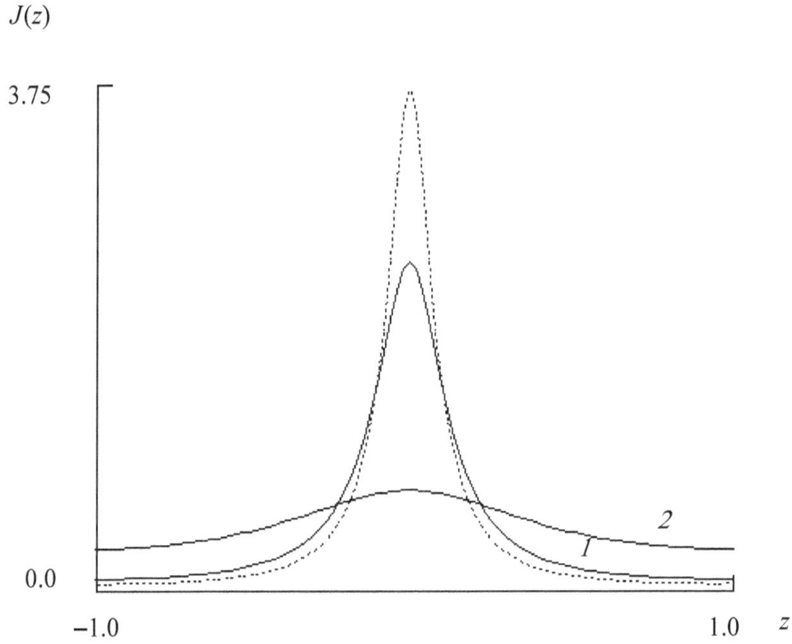

Fig. 6.3. Profile of the diffraction line of crystals with a correlated distribution of dislocations loops in the slip plane: (*1*) $\tau = b$; (*2*) $\tau = 3b$ ($\xi = 500\ a$). Dashed line for chaotic distribution of loops

Effect of correlation in the plane of the dislocations loops is more significant than along the normal to it. An increase in the parameter μ modifies the curve $T_1(m)$ with m close to zero. An increase in the parameter τ sharply rises the rate of growth $T_1(m)$.

If the diffraction line is observable when the dislocation density in the crystal is about 10^{10} cm^{-2}, then the correlation radius τ can not exceed $\sim 3b$, where b is the interatomic distance, otherwise the line would be smeared into a diffuse background, as it is seen by Fig. 6.3.

Existence of large correlation radii in the slip plane is not realistic. Correlation radius τ predetermines the density of pile-ups of loops. Range of far bond of loops (and their pile-ups) in the elastic stresses field corresponds to the stability of the system as a whole.

4. Models of long-range order in dislocation systems. Diffraction effects from the long-range ordering in the system of dislocations are considered on two models of periodic structures.

In the first model dislocation loops are placed exactly along parallel slip planes with a constant step ℓ in the directions of $\mathbf{j}=\sqrt{\frac{1}{2}}\langle 110\rangle$. In the second model loops of dislocations form the networks with the translation period $\ell = 2\xi$ in two directions $\mathbf{j}=\sqrt{\frac{1}{3}}\langle 111\rangle$ in the slip planes. There are taken the sizes of all loops to be $\xi = \overline{\xi}$, and degree of order $\overline{\eta} = 1$.

Periodic component T_2 is calculated for the reciprocal lattice with the basis vector **g**, the module of projection of which on the plane of the dislocations loops is $g' = (2\pi/\ell)\sqrt{1-(\mathbf{n}\cdot\mathbf{j})^2}$.

Such an expression should be in the T_2 Eq. (6.13)

$$\Theta_{\alpha\beta} = \sum_{\mathbf{j}} \gamma_{ij}^{(\alpha\beta)} \mathrm{K} \left\langle \left(\frac{a}{\ell}\right)^3 \left[1 - \cos\frac{2\pi}{\ell}(\mathbf{jR})\right] \Big/ \left(\frac{2\pi}{\ell}\right)^2 \delta_{\alpha\beta} \right\rangle_{1/\ell}, \quad (6.16)$$

$$\mathrm{K} = \begin{cases} \frac{1}{4} & \text{for waves along normal } \mathbf{n}, \\ \left[\frac{J_1(\pi)}{\pi}\right]^2 & \text{for networks in the plane } \mathbf{n}. \end{cases}$$

Summation is carried out over the basis vectors **g** in the reciprocal lattice of the ordering of defects of each type α.

In the considered examples, the period of the reciprocal lattice $\left(\frac{2\pi}{\ell}\right)$ is considered to be equal to the expected value in the ensemble of dislocation systems, it is supposedly the same for all types of loops ($\alpha = 1, \ldots, p$).

The form of the averaged periodic component T_2 for ordering along the normal to the slip planes that creates the concentration waves of different lengths is shown in Fig. 6.4.

Given an ideal periodicity of loops placement along planes with interval $\ell < \overline{\xi}$ the diffraction line would disappear in turbulent ($\overline{\xi} \leq 100\,a$) or in quiet ($\overline{\xi} \geq 250\,a$) waves. Under the concentration fluctuations with a period $\ell > \overline{\xi}$ when small dislocations loops the diffraction line is weaken, and at large loops is broaden, approaching by shape to Gaussian (Fig. 6.5).

In crystals with a high concentration of large dislocation loops the one-dimensional periodicities, if any occur, then in insignificant shares $\overline{\eta}$[1].

---

[1] What difficulties are created by slip bands for measuring the average dislocation density in crystals are discussed in Ref. [104].

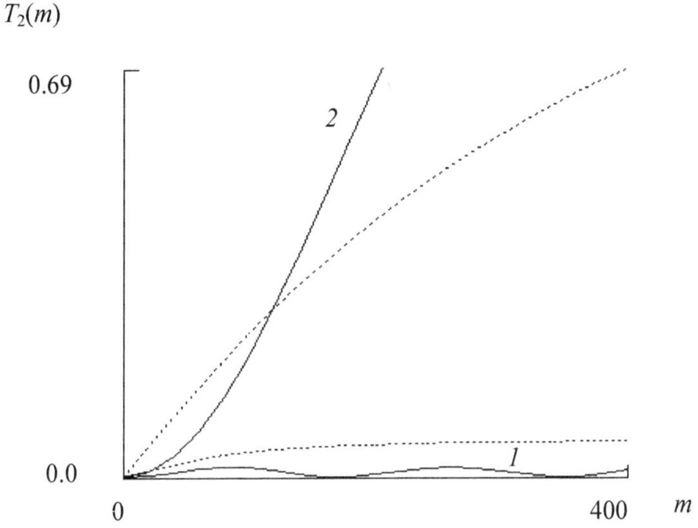

Fig. 6.4. Periodic component of the logarithm of harmonics when dislocation loops are placed with the given step along the normal to the slip plane: (*1*) $\xi = 25\ a$, $\ell = 50\ d$; (*2*) $\xi = 100\ a$, $\ell = 200\ d$. Dashed lines show the component $T_0(m)$

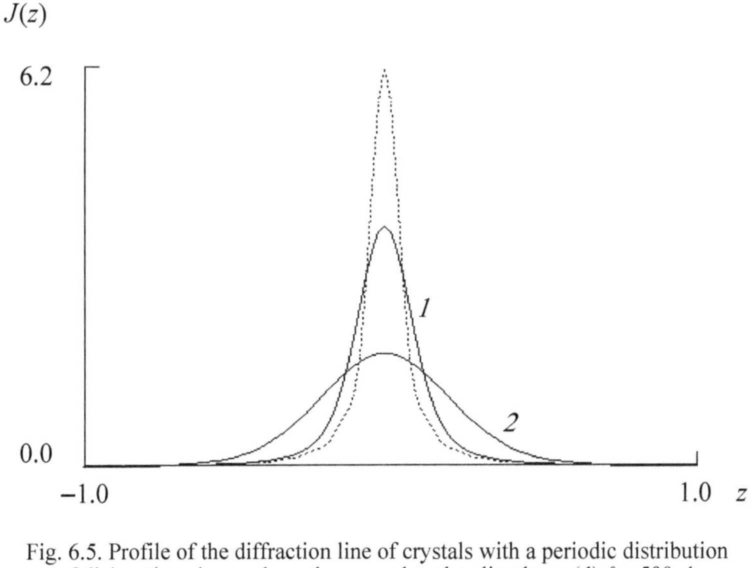

Fig. 6.5. Profile of the diffraction line of crystals with a periodic distribution of dislocations loops along the normal to the slip plane: (*1*) $\ell = 500\ d$; (*2*) $\ell = 250\ d$ ($\xi = 250\ a$). Dashed line for chaotic distribution of loops

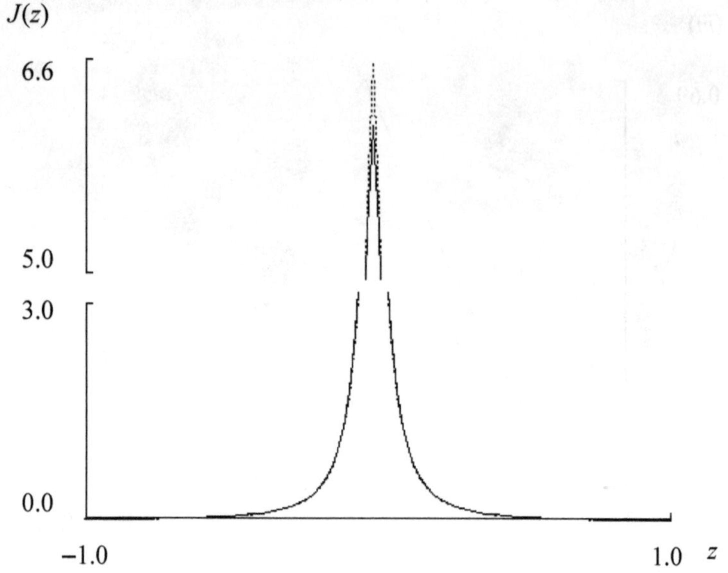

Fig. 6.6. Profile of the diffraction line of crystals with the plane networks of dislocations loops ($\xi = 250\ a$). Dashed lines for chaotic loops distribution

Ordering of the dislocation loops with the formation of plane networks (sub-boundaries), accompanied by the appearance of the periodic component T_2, leads only to an insignificant weakening of the diffraction line, as shown in Fig. 6.6.

Transition of dislocations from chaotic distribution in volume to regular networks randomly distributed over crystallographic planes exerts little effect on the diffraction intensity distribution (as well in the case of the organization of walls from rectilinear dislocations [41]).

5. Factor of the crystals size. When a small size of the scattering crystals the sine Fourier coefficients of the diffraction line appear the relative magnitude of which β, as shown in Eq. (6.10), is related to the Fourier transform of the shape function of a crystal $s_{\mathbf{k}}$.

If taken crystal in a plate shape of thickness Ξ in the direction of normal to the reflecting plane, then

$$\begin{cases} |s_{\mathbf{k}}|^2 = N^2 \delta(k_x)\delta(k_y)\psi(k_z,\Xi), \\ \psi(k_z,\Xi) = \dfrac{\sin^2(k_z \Xi/2)}{(k_z \Xi/2)^2}. \end{cases}$$

When calculating $\beta \equiv \beta(\mathbf{q}, \mathbf{R})$ from Eq. (6.10) with a given function $|s_\mathbf{k}|^2$ the sum over **k**-space reduces to a one-dimensional sum over k_z, and it can be replaced by integral multiplied by the normalization constant $\Xi/2\pi$. Since under the integral it is appeared the function $h_z^\alpha(k_z)$ from Eq. (6.9), which is odd in k_z, then in the chosen coordinate system the quantity under review takes the form

$$\beta_m = -2m\bar{c}(\xi/a)^2(\Xi/a)\int_0^\infty Y(k_z,\xi)\psi(k_z,\Xi)\frac{\sin(k_z R_z)}{k_z R_z}\,dk_z,$$

$$\begin{cases} Y(k_z,\xi) = \dfrac{1}{p}\displaystyle\sum_{\alpha=1}^{p} \mathrm{N}_{zz}^\alpha\left[2\dfrac{J_1\left(k_z\sqrt{1-|n_z^\alpha|^2}\,\xi\right)}{k_z\sqrt{1-|n_z^\alpha|^2}\,\xi} - 1\right], \\ \mathrm{N}_{zz}^\alpha = \dfrac{2\pi}{a}\left(b_z^\alpha n_z^\alpha\right)(1-\sigma_1). \end{cases}$$

Function in the square brackets of the formula $Y(k_z, \xi)$ is equal to zero when $k_z = 0$ and approaches unity with a minus sign when $|k_z| \to \infty$.

Figure 6.7 gives a represent on effect of the thickness of scattering crystal and the size of dislocation loops on a value of the sinus harmonics of the diffraction line.

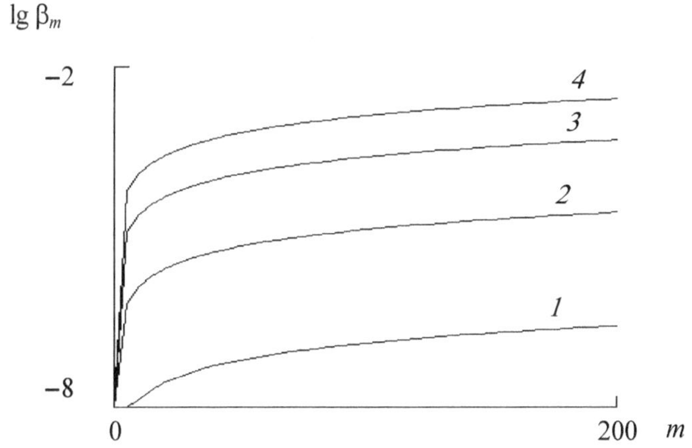

Fig. 6.7. Relative value of the sinus component of harmonics of the diffraction line in dependence on thickness of scattering crystal: (*1*, *3*) $(\Xi/2\xi) = 100$; (*2*, *4*) $(\Xi/2\xi) = 10$; (*1*, *2*) $\xi = 50\,a$; (*3*, *4*) $\xi = 500\,a$

While the thickness of a crystal is not too small ($\Xi > 1$ μm), the size of the dislocation loops affects the value of the sine harmonics more strongly: with increasing ξ by an order of magnitude, β increases by three orders of magnitude. Decrease in the relative thickness of a scattering crystal $\Xi/2\xi$ becomes a significant factor with the small dislocations loops; in that case growth of β goes ten times quicker.

§ 6.3. Observability of the Dislocations System in the Diffraction Space

General diffraction equation is splitting into simple expressions with the finding of measurable states of the dislocation structure of crystals.

1. Mapping of the states of a dislocation structure in the diffraction space. Equation of harmonics of the diffraction line contains the sum of the integrals Z (6.14) of the cylindrical Bessel functions ($J_0(x), J_1(x)$).

Let us introduce the representation

$$Z = \xi^{-2}\int_0^\infty \varphi_1(t)\,\varphi_2(t)\,t\,dt,$$

$$\varphi_1(t) = \left[J_1(\xi t)\right]^2 / t,\quad \varphi_2(t) = \left[1 - e^{-ut} J_0(vt)\right] / t^2,$$

in which the functions $\varphi_1(t)$, $\varphi_2(t)$ have the integral Hankel transform [92]

$$\psi(s) = \int_0^\infty \varphi(t) J_1(st)\, t\, dt:$$

$$\begin{cases} \psi_1(s) = \begin{cases} (2\pi)^{-1}\xi^{-2}\sqrt{(2\xi)^2 - s^2}, & 0 < s < 2\xi, \\ 0, & 2\xi < s < \infty; \end{cases} \\ \psi_2(s) = 1 - s\left[\dfrac{1}{\pi}\displaystyle\int_0^\pi \sin^2 x\, dx \Big/ \sqrt{(u + is\cos x)^2 + v^2}\right] \quad (u, v > 0). \end{cases}$$

Due to Parseval equality [37]:

$$\int_0^\infty \varphi_1(t)\varphi_2(t)\, t\, dt = \int_0^\infty \psi_1(s)\psi_2(s)\, s\, ds,$$

we obtain computing formula

$$Z = (2\pi)^{-1}\xi^{-4}\int_0^{2\xi}\sqrt{(2\xi)^2 - s^2}\, y(s)\, s\, ds.$$

Here, $y(s)$ denotes the elementary functions to which the $\psi_2(s)$ is reduced:

$$y(s) = \begin{cases} 1 - \dfrac{1}{2}R^{-1}s, & s \ll R, \\ (\mathbf{R}\cdot\mathbf{n})/s, & s \gg R. \end{cases}$$

Formally, $y(s)$ is the Laurent series [37], which does not have a common region of convergence of its two parts: expansions in positive and negative powers of the variable s, where the second-order terms are made vanish.

Eventually

$$Z = \begin{cases} \dfrac{4}{3\pi}\xi^{-1} - \dfrac{1}{4}R^{-1}, & 2\xi \ll R, \\ \dfrac{1}{2}\xi^{-2}(\mathbf{R}\cdot\mathbf{n}), & 2\xi \gg R. \end{cases}$$

It follows that the state in the field of small dislocation loops is measurable over higher harmonics of the diffraction line, and in the field of large loop sizes it is measurable over the initial harmonics. In the former case, the optimal region of observations is the central part of the diffraction line with the small $\{HKL\}$, in the latter case it is "the wings endings" of the line with the large $\{HKL\}$.

For observations in the intrinsic regions of the diffraction space there is a simple mathematical expression:

$$T \cong \begin{cases} \bar{c}(\xi/a)^3 C_0\, Q_{HKL}^2\left[\dfrac{8}{3\pi} - \dfrac{1}{2}(\xi/a)Q_{HKL}m^{-1}\right], & \\ & m/Q_{HKL} > (2\xi/a), \\ \bar{c}(\xi/a)^2 C_1\, Q_{HKL}\, m, & m/Q_{HKL} < (2\xi/a); \end{cases} \qquad (6.17)$$

$$Q_{HKL} = \sqrt{H^2 + K^2 + L^2}\ ;$$

(C_0, C_1) are crystallo-geometric constants for calculation of scattering by a polycrystal that contain the tensor coefficients $\Gamma_{ij}^{(\alpha\beta)}$ for all available slip systems:

$\langle 111\rangle\{110\}$ bcc: $C_0 = 16.278295$, $C_1 = 8.769899$;
$\langle 111\rangle\{112\}$ bcc: $C_0 = 16.278295$, $C_1 = 9.493694$;
$\langle 110\rangle\{111\}$ fcc: $C_0 = 10.852197$, $C_1 = 6.265519$.

Figure 6.8 on the example of the model structure of bcc crystals demonstrates intermediate state at an intermediate level of the dislocation loops sizes.

Intermediate state is observable in a very limited number of harmonics of the order $k = æ m$ ($k = 1, 2, \ldots$) where æ is the ratio of the interval of the diffraction line measurement to the period of the reciprocal lattice of crystal. Therefore, the parameters of the dislocations system will have to be determined over two equations of the model with different Q_{HKL}.

2. Distinct correlation effects in the distribution of dislocation loops. Equation (6.17) represents chaotic distribution of dislocation loops ($T \cong T_0$). Experience reveals the existence of local fluctuations in concentration of loops inherent to short-range order.

Let us accept a correlation model describing the short-range order, as in § 6.2. Parameters of Eq. (6.15) are subject to the relation $\tau, \mu \ll R$, as in reality the correlation is short-range, and small R are unavailable for observing since $R \sim k/æ$ where $æ \ll 1$. Under these conditions it can be used the approximation to the correlation component Eq. (6.12), (6.15):

$$T_1 \cong q_i q_j \bar{c}(a/2\pi)^3 (\xi/a)^4 \left[\frac{1}{p}\sum_{\alpha=1}^{p} \Gamma_{ij}^{(\alpha\beta)} 2\pi^2 (\upsilon \mathrm{E})\delta_{\alpha\beta}\right],$$

$$\begin{cases} \mathrm{E} = \mathrm{Z} - \mu \displaystyle\int_0^\infty \left[J_1(\xi t)/(\xi t)\right]^2 t\,dt = \mathrm{Z} - (2\xi)^{-1}(\mu/\xi), \\ \upsilon = \dfrac{2}{p}\left[(\mu/d)(\tau/b)^2\right]. \end{cases}$$

Resulting equation with the parameter υ proportional to volume of correlation ellipsoid is reduced to the form

$$T_1 \cong \upsilon[T_0 - \Upsilon], \quad \Upsilon = (\mu/\xi)\,\bar{c}\,(\xi/a)^3 C_0 Q_{HKL}^2.$$

Function Υ arising in the three-dimensional correlation model does not depend on m. An approximation of T_0 it is Eq. (6.17).

Significance of the factor of space correlation under small and large dislocation loops revealed itself on examples of bcc crystals with specified parameters, represented in Fig. 6.9.

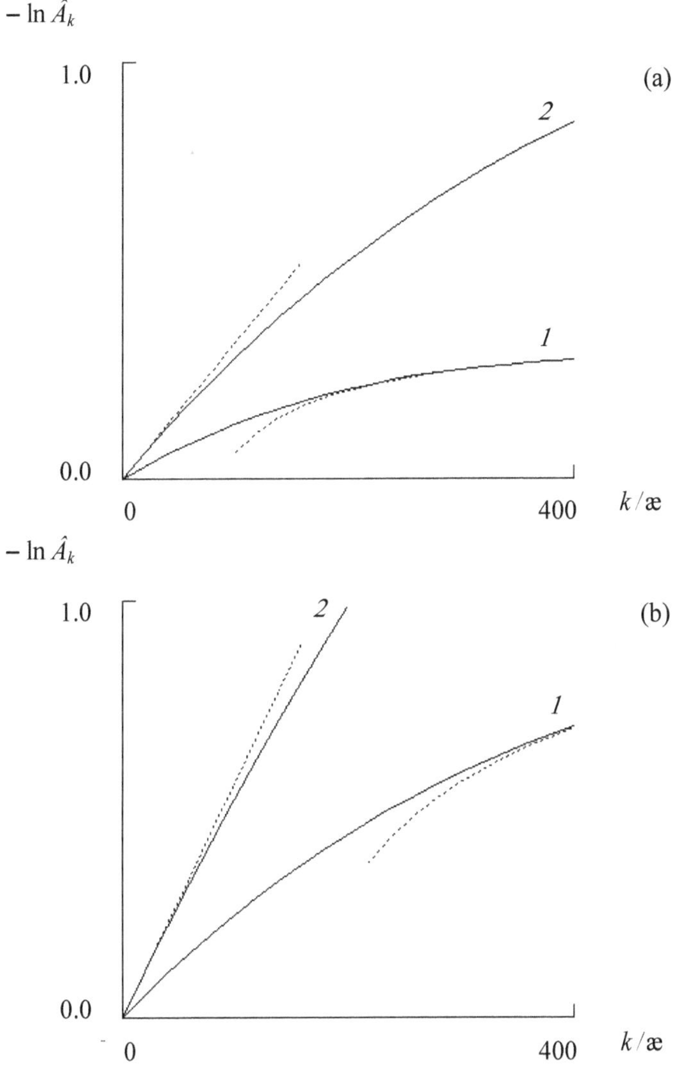

Fig. 6.8. Regions of limitations for the model of diffraction observations (a) {110} and (b) {112} when an intermediate level of the dislocation loops sizes: (*1*) $\xi = 100\ a$; (*2*) $\xi = 200\ a$. Dashed line indicates the increasing deviations from the exact theoretical equation

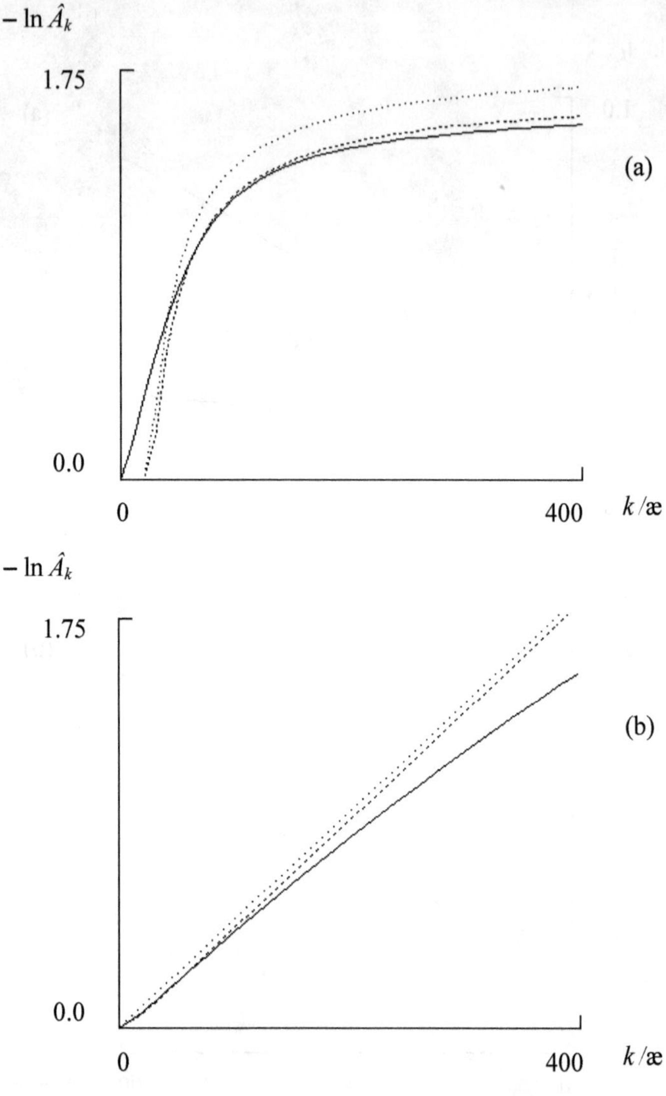

Fig. 6.9. Errors of the model of diffraction observations upon correlation in the system of dislocation loops ($\tau/b = 3$, $\mu/d = 3$): (a) {110} ($\xi = 30\ a$, $\rho_d = 10^{11}$ cm^{-2}) and (b) {112} ($\xi = 300\ a$, $\rho_d = 10^{9}$ cm^{-2}). Dashed and dotted lines, respectively, three-dimensional and two-dimensional approximation

Spatial form of the correlation for dislocation loops of large sizes $(\tau, \mu \ll \xi)$ is practically indistinguishable from the correlation in its plane with an equivalent parameter

$$\upsilon = (2\pi/p)\left(\bar{\tau}/b\right)^2, \quad \bar{\tau} = \tau\sqrt{\pi^{-1}(\mu/d)};$$

$\bar{\tau}$ is interpreted as an effective radius of correlation in the slip plane.

Correlation in the distribution of large dislocation loops is equivalent to an increase in their concentration by approximately $\left[1 + (2\pi/p)\left(\bar{\tau}/b\right)^2\right]$ times. The smaller the number of slip systems *p*, the more significant the role of correlation.

3. Observable long-range order in the system of dislocation loops. Let there be correlation in the system of dislocations and long-range order in the arrangement of loops on the normal to the slip planes $\mathbf{n}^\alpha$ ($\alpha = 1, \ldots, p$). The proportion of ordered loops of their total number in a crystal is a degree of order $\bar{\eta}$.

A periodic component will appear in the diffraction equation:

$$T = T_0 + \left(1 - \bar{\eta}\right)T_1 + T_2.$$

Component T_2 (6.13) is calculated under the assumption that the distribution of loop sizes ξ and the distribution of order period ℓ in the statistical ensemble of dislocation systems are mutually independent and the same for loops of different type ($\alpha = 1, \ldots, p$).

The one-dimensional reciprocal lattice of ordered dislocations in a crystal of finite size Ξ can be represented as a periodic wave packet with a limited spectrum. In the reciprocal space peaks with spacing $\left\langle 1/\ell \right\rangle$ and width $\Delta \sim \left(1/\Xi\right)$ will appear. Angle brackets denote the expected values of parameters in the statistical ensemble of dislocation systems.

Within restrictions area of the theory, when period of inhomogeneity of crystal distortions is much smaller than its size, variance of the distribution $\sigma^2_{1/\ell}$ is subject to the relation $\sigma_{1/\ell}/\left\langle 1/\ell \right\rangle < \left(\bar{\ell}/\Xi\right) < 0.1$. Here, it is taken into account that $\left\langle 1/\ell \right\rangle > 1/\langle \ell \rangle$, and the expected $\langle \ell \rangle$ in the ensemble coincide with the average $\bar{\ell}$ over the crystal volume (according to the ergodic hypothesis [44]).

The approximate equation of the periodic component applicable for determining the order parameters will be obtained using formula (6.16) under asumption $\frac{2\pi}{\ell}(\mathbf{jR}) \ll 1$, where $|\mathbf{R}| = m\,a/Q_{HKL}$. When for each loops type of α there is only one translation vector $\mathbf{j} \parallel \mathbf{n}^{\alpha}$ ($\alpha = 1, \ldots, p$), the equation for periodic component takes the form:

$$T_2 \cong \overline{c}\,\overline{\eta}(\xi/a)^4\left[\tfrac{1}{4}(a/\ell)^3 C_2\, m^2\right],\ m/Q_{HKL} < (\ell/a)/2\pi. \quad (6.18)$$

To calculate diffraction on polycrystals with long-range order in the dislocation structure, it is added the crystal-geometrical coefficient C_2:

⟨111⟩ {110} bcc:	$C_2 = 5.411616$;
⟨111⟩ {112} bcc:	$C_2 = 5.411616$;
⟨110⟩ {111} fcc:	$C_2 = 3.607744$.

Figure 6.10 gives presentation of the effect of long-range order on harmonics of the diffraction line when different parameters of the dislocation system. Increasing errors of the approximate equation with measurable of the order parameters are visible in dashed lines.

The area of observability of long-range order in the dislocation structure of crystals is significantly limited in the diffraction space. With large dislocation loops in crystals, the long-range order is practically unavailable for observation because of the suppressing effect of correlation, what is verified by simulation tests on model crystals, which are used for the examples given in Ch. 7.

4. Diffraction model under conditions of limited observability of states of the dislocation structure. Averaging of the approximate function $T = T_0 + T_1$ by logarithmically normal distribution of ξ is reduced to determining the moments of this distribution [25]. (Deviations from randomness, when only a small fraction of loops sometimes forms regular networks, can be neglected.)

Equations of the model of diffraction observations for real crystals take the form

$$T \cong \begin{cases} \overline{c}\left(\overline{\xi}/a\right)^4\left[1+\left(\sigma_\xi/\overline{\xi}\right)^2\right]^6\left[1+(2\pi/p)\left(\overline{\tau}/b\right)^2\right]\times \\ \times\left(-\frac{1}{2}C_0\,Q_{HKL}^3 m^{-1}\right) + L_{HKL},\quad m/Q_{HKL} > 2\overline{\xi}/a, \\ \\ \overline{c}\left(\overline{\xi}/a\right)^2\left[1+\left(\sigma_\xi/\overline{\xi}\right)^2\right]\left[1+(2\pi/p)\left(\overline{\tau}/b\right)^2\right]\times \\ \times\left(C_1\,Q_{HKL}\,m\right) + M_{HKL},\quad m/Q_{HKL} < 2\overline{\xi}/a. \end{cases} \quad (6.19)$$

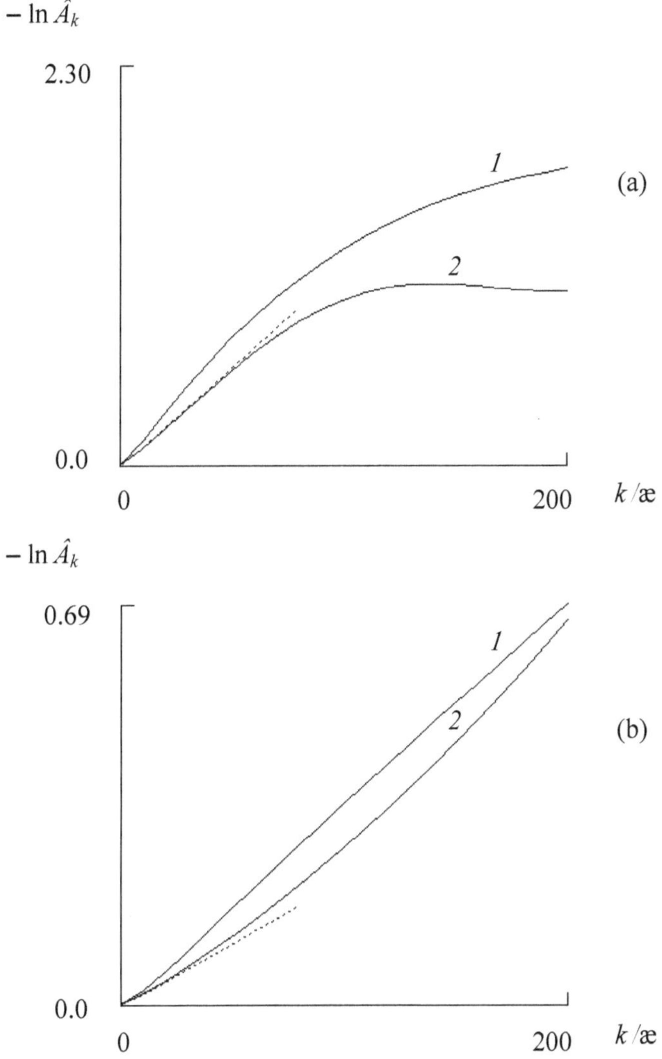

Fig. 6.10. Diffraction line harmonics behavior when long-range order in the dislocation system:
(a) {112} bcc ($\xi/a = 30$, $\rho_d = 10^{11}$ cm^{-2});
(b) {113} fcc ($\xi/a = 300$, $\rho_d = 10^{9}$ cm^{-2});
(*1*) $\bar{\eta} = 0$, (*2*) $\bar{\eta} = 0.75$ ($\ell/\xi = 2$, $\tau/b = 2$, $\mu/d = 2$)

Here, $\left(\bar{\xi}/a\right)$ and $\left(\sigma_{\xi}/\bar{\xi}\right)$ are the average normalized radius of dislocation loops and the coefficient of variation of their sizes.

The terms L_{HKL}, M_{HKL} independent of m include an additional correlation parameter $\left(\mu/\bar{\xi}\right)$. Their explicit expressions are given in [73].

By the equations of the model it is clear that the diffraction pattern is differ in sensitive to quantitative changes in the sizes of dislocation loops when they are of different qualitative levels. The study of the model reveals that for large dislocation loops there is a high sensitivity to correlation in the loop plane, and for small loops to correlation along normal to it.

The effect of a long-range order in the system of dislocation loops, which is revealed by the averaged periodic component T_2 (6.18) for crystals of finite size, is considered in § 7.3.

Measuring interval in diffraction space: Coefficient æ denoting the ratio of observable diffraction spacing to the theoretical interval Fourier representation of diffraction line is important in subsequent determination of parameters of the dislocation structure. Herein it is given a strictly mathematical derivation of the formula to calculate correctly of æ.

According to the geometry of experiment the length of the diffraction vector for crystallographic planes of type $\{HKL\}$ has the formula [10]

$$q = 4\pi\frac{\sin\theta_{HKL}}{\lambda}, \qquad dq = 2\pi\frac{\cos\theta^{0}_{HKL}}{\lambda}d\left(2\theta_{HKL}\right),$$

$$-\frac{1}{2}\Delta\left(2\theta_{HKL}\right) \le 2\left(\theta_{HKL} - \theta^{0}_{HKL}\right) \le \frac{1}{2}\Delta\left(2\theta_{HKL}\right),$$

where $\Delta\left(2\theta_{HKL}\right)$ is the interval to be measured of scattering angles with center of $2\theta^{0}_{HKL}$, and λ is the wavelength of the radiation.

Changing in the length of the diffraction vector near the reciprocal lattice site corresponds to small deviations $\{HKL\}$ from their integer values in the reciprocal space of perfect crystal. In the Cartesian coordinate system (X, Y, Z)

$$dq = \sqrt{dq_X^2 + dq_Y^2 + dq_Z^2}.$$

The coordinate system is constructed so that

$$\mathbf{q} = \frac{2\pi}{a}\left[0, 0, Q_{HKL}\right], \; Q_{HKL} = \sqrt{H^2 + K^2 + L^2},$$

$$dq = \frac{2\pi}{a}\sqrt{\left(\frac{\partial Q_{HKL}}{\partial H}\right)^2 dH^2 + \left(\frac{\partial Q_{HKL}}{\partial K}\right)^2 dK^2 + \left(\frac{\partial Q_{HKL}}{\partial L}\right)^2 dL^2},$$

$$dH = dK = dL = dt,$$

$$dq = \frac{2\pi}{a}dt, \ -\frac{æ}{2} \le t \le \frac{æ}{2} \quad (0 < æ < 1).$$

Experiment produces a diffraction intensity distribution along the length of the vector

$$\int dq = 2\pi \frac{\cos\theta^0_{HKL}}{\lambda}\Delta(2\theta_{HKL}) = \frac{2\pi}{a}æ.$$

From the equality obtained it follows that

$$æ = a\frac{\cos\theta^0_{HKL}}{\lambda}\Delta(2\theta_{HKL}).$$

Precisely this is the measured part of the reciprocal lattice period of crystal $\frac{2\pi}{a}$.

Observed interval under tetragonal distortions of crystals: For crystals with unit cell dimensions (a, c) and interplanar spacing d_{HKL} [98]

$$\frac{1}{d_{HKL}} = \sqrt{\frac{H^2 + K^2}{a^2} + \frac{L^2}{c^2}}$$

the length of the diffraction vector is

$$q = \frac{2\pi}{d_{HKL}} = \frac{2\pi}{a}Q_{HKL}, \ Q_{HKL} = \sqrt{H^2 + K^2 + \left(\frac{a}{c}\right)^2 L^2}.$$

With $\{HKL\}$ deviation from the integer values by the same part of the reciprocal lattice period it is follows

$$\frac{dH}{a} = \frac{dK}{a} = \frac{dL}{c}, \ dH = dK = \frac{a}{c}dL = dt,$$

$$dQ_{HKL} = dt, \ dq = \frac{2\pi}{a}dt.$$

As a result for crystals with tetragonality extent $\frac{c}{a} > 1$ for the same diffraction angles interval of $\Delta(2\theta_{HKL})$ an observed part of the theoretical interval is $\frac{a}{a_0}$ times smaller than for cubic crystals with lattice period a_0.

CHAPTER 7
ESTIMATION OF PARAMETERS OF THE DISLOCATION SYSTEM BY MEASURED HARMONICS OF THE DIFFRACTION LINE

Equations of the diffraction theory became the basis of methods for identifying a system of dislocation loops model a real structure of deformed crystals [72–73].

§ 7.1. Identification of a Random System of Dislocations in Strongly Distorted Crystals

A system of dislocations loops is identifiable by means of the model of diffraction observations that corresponds to state of the object structure.

1. Assessment of the structure state characterized by the level of dislocation loops sizes. Equation of normalized harmonics of the diffraction line

$$A_k \cong e^{-T(m)},$$

which have the order of $k = æ\, m$ in the interval of measurement æ from period of the reciprocal crystal lattice, can be represented in the form of a regression model of observations $Y_k = -\ln A_k$ ($k = 1, 2, \ldots$):

$$Y_k \cong \mathbf{z}_k \mathbf{h}. \tag{7.1}$$

Here, $\mathbf{h} = (h_0, h_1)^{\mathrm{t}}$ is the vector of the regression coefficients (t-superscript denotes transpose), and $\mathbf{z}_k = (z_{k0}, z_{k1})$ is the vector of the basis functions of regression, which corresponds to the domain of definition of $T(m)$ Eq. (6.19):

$$\left.\begin{array}{l} z_{k0} = C_0 Q_{HKL}^2,\; z_{k1} = -\dfrac{1}{2} C_0 Q_{HKL}^3 (k/æ)^{-1} \\ \left[(k_{\min}/æ)/Q_{HKL} > 2\bar{\xi}/a\right]; \end{array}\right\} \tag{7.2}$$

$$\left.\begin{array}{ll} z_{k0} = -C_0 Q_{HKL}^2, & z_{k1} = C_1 Q_{HKL} (k/æ) \\ \left[(k_{\max}/æ)/Q_{HKL} < 2\bar{\xi}/a\right]. & \end{array}\right\} \tag{7.3}$$

On the equation of the best agreement with the data is recognized the state of dislocation structure. Choose of the equation is realizable by sequential regression analysis of the observations Y_k:

(*1*) By the method of successive including in the observation vector $(Y_1, \ldots, Y_k)$ of higher order elements Y_k ($k = 2, 3, \ldots$) it is searched the largest portion of observation interval $[1, k]$, where the equation of approximation of the initial harmonics is adequate in the basis (7.3). If this equation is rejected with a high probability P already at small k, then there should go to the definition of $k_{\min}$ for the equation of approximation of the highest harmonics in the basis (7.2).

(*2*) By the method of successive excluding from the observation vector $\left(Y_k, \ldots, Y_{k_{\max}}\right)$ of lower order elements Y_k ($k = 1, 2, \ldots$) it is searched the off cut $[k, k_{\max}]$, in which the regression equation is adequate in the basis (7.2), and the ratio of its coefficients h_1/h_0 is well-defined and reaches a maximum value. Simultaneously, since $Y_k > 0$, $\left|(k_{\max}/æ)/Q_{HKL} - 2(h_1/h_0)\right|$ is minimized, and k approaches the allowed limit $k_{\min}$. An estimate of the vector $\mathbf{h}^* = \left(h_0^*, h_1^*\right)^{\mathrm{t}}$ gives a limitation on both the average loops sizes and the dislocation density: $\bar{\xi}/a < h_1^*/h_0^*$; $\rho_d < h_0^*\left(h_1^*/h_0^*\right)^{-2}\left(2\pi/a^2\right)$.

If there is no reasons for rejecting the regression equation in the basis (7.3) for the entire interval $1 \le k \le k_{\max}$, where $k_{\max}$ is the largest order of the reliable harmonics A_k with the mean square error of measurements σ_{A_k}, then $2\left(\bar{\xi}/a\right) > (k_{\max}/æ)/Q_{HKL}$, thus the data suggest that the dislocation loops in crystals are large, and the average dislocations density $\rho_d \ll 2h_1^*\left[(k_{\max}/æ)/Q_{HKL}\right]^{-1}\left(2\pi/a^2\right)$. The coefficient h_0^*, when it is well defined, will show the lower estimate of the correlation parameter $\mu/d > h_0^*/h_1^*$.

The equation is rejected with reliability not less than P if the weighted (with a weight $A_k^2/\sigma_{A_k}^2$) sum of $N = k_{\max} - k_{\min} + 1$ squares of residual deviations exceeds $(N - n)/(1 - P)$, where n is the dimension of the regression coefficients vector [2].

Procedure for discriminating the equations of the diffraction observations model performs a nonparametric identification the system of dislocations.

Systematic errors of the observations: The main systematic error of the data A_k arises from the splitting of X-ray line, which is a doublet of the $K_{\alpha_1} - K_{\alpha_2}$ radiation. Further it is shown that the error grows as k^2.

Fourier coefficients of the diffraction doublet receive the expression

$$A_k^d = A_k^s\left[w_1 + w_2\cos(k\beta)\right], \quad B_k^d = A_k^s\left[w_2\sin(k\beta)\right].$$

Here, β is the ratio of the inter-doublet spacing to the range of Fourier series expansion, and w_1 and w_2 are the weight fractions of doublet components whose profile is described by the harmonics A_k^s. A center of the expansion is chosen by a maximum of the diffraction intensity.

Subject to that $\beta \ll 1$,

$$\left|A_k^d\right|^2 + \left|B_k^d\right|^2 \cong \left|A_k^s\right|^2 \left[1 - (w_1 w_2 \beta^2) k^2\right].$$

Consequently, the observations that available by determining the physical profile of the diffraction line are of form

$$\ln A_k \cong \ln \left(A_k^s / A_0^s\right) - \frac{1}{2}\left(w_1 w_2 \beta^2\right) k^2.$$

A shift of a center of the Fourier series expansion does not affect the result.

There is a method of separating the lines of the diffraction spectrum with an optimal estimation of the Fourier coefficients A_k^s [69–70]. Such way of correcting the data A_k would be too costly. The best solution is to use the available information on systematic errors.

2. Adaptive model for parametric identification of the dislocations system. Parametric identification is complicated by the presence of systematic errors of the harmonics of the diffraction line that grow with increasing of their order. Model of diffraction observations in the form allowing estimation of parameters should include information on regular deviations of the data.

Let $\boldsymbol{\theta}$ be the vector of the dislocation structure parameters to be estimated, x_k be an independent variable, and u_k be the auxiliary variable compensating for the systematic errors of the original model and data. In the absence of errors, the vector of functions of the model is subject to equality

$$\left.\begin{array}{l} \mathbf{g}(\mathbf{A}, \boldsymbol{\theta}) = 0, \\ \mathbf{g} = \left\{g_k\left(A_k, \boldsymbol{\theta}\right)\right\}, \; g_k = \ln A_k + x_k \prod_{q=1}^{4} e^{\theta_q} + u_k. \end{array}\right\} \quad (7.4)$$

The model equations appropriate to different states of the dislocation structure differ by expressions of variables and parameters:

$$\left.\begin{array}{l} x_k = -\frac{1}{2} C_0 Q_{HKL}^3 (k/æ)^{-1} \quad \left(k_{\min} \le k \le k_{\max}\right), \\ \left\{\begin{array}{l} \theta_2 = 3 \ln\left(\xi/a\right), \\ \theta_3 = 6 \ln\left[1 + \left(\sigma_\xi/\xi\right)^2\right] \end{array}\right. \end{array}\right\} \text{for small loops;}$$

$$\left.\begin{array}{l} x_k = C_1\, Q_{HKL}\left(k/æ\right) \quad \left(1 \le k \le k_{\max}\right), \\ \begin{cases} \theta_2 = \ln\left(\bar{\xi}/a\right), \\ \theta_3 = \ln\left[1+\left(\sigma_\xi/\bar{\xi}\right)^2\right] \end{cases} \end{array}\right\} \text{for large loops.}$$

The components of vector $\boldsymbol{\theta}$

$$\begin{cases} \theta_1 = \ln\left[\bar{c}\left(\bar{\xi}/a\right)\right], \\ \theta_4 = \ln\left[1+(2\pi/p)\left(\bar{\tau}/b\right)^2\right]. \end{cases}$$

are the same in two equations of the model. (Parameter $\mu/\bar{\xi}$ is related to $\boldsymbol{\theta}$ by the equation given in [73].)

In the vector of auxiliary variables $\mathbf{u}$ the components u_k $(k > 0)$ have the term u_0 that makes up for not depend on k members of diffraction model in Eq. (6.17). For large dislocation loops a value u_0 is negligibly small due to a weakly exhibited effect of spatial correlation (§ 6.3).

Search for a missing information to accurately describe the real observations over the object goes in the process of identifying the model of object.

When a non-parametric identification into the vector of basis functions $\mathbf{z}_k$ (7.2) – (7.3) it is introduced component $z_{k2} = \left(k/æ\right)^2$. Supplementary regression coefficient h_2^* appearing in $\mathbf{h}^*$ is suitable for modeling the measured vector $\mathbf{u}$. It is assumed that the variables u_k contained therein are independent normally distributed random variables with the mathematical expectation and variance of

$$\bar{u}_k \approx z_{k2}\, h_2^* + z_{k0}\, h_0^*, \quad \sigma_{u_k}^2 \approx z_{k2}^2\, \sigma_{h_2}^2 + z_{k0}^2\, \sigma_{h_0}^2 .$$

Measured vector of an auxiliary variables $\mathbf{u}$ performs the adjustment of the model in Eq. (7.4), through which parametric identification is made, refined itself in concert with the estimate of the parameters vector $\boldsymbol{\theta}$ and approaching actual systematic errors $\hat{\mathbf{u}}$.

There exists an infinite set of vectors $\boldsymbol{\theta}$ satisfying condition

$$\left|\prod_{q=1}^{4} e^{\theta_q} - h_1^*\right|^2 \le \sigma_{h_1}^2,$$

for which fitting the model Eq. (7.4) is not worse than regression Eq. (7.1). An estimate, representing the physical reality, can be found only with the involvement of the information about the allowability of the quantities entering the vector $\boldsymbol{\theta}$.

The information on the possible values of the dislocation density and the average sizes of the loops, experience provides. The order of value of the correlation parameter in the plane of dislocation loops is came up by theoretical calculations (§ 6.2).

3. Nontraditional method of statistical estimation of parameters. Let the results of the experiment are represented in the K-dimensional sample space ($K \equiv k_{max}$, when $k_{min} = 1$). A vector of diffraction observations $\mathbf{A} = (A_1, \dots, A_K)$ determines a point of the sample space. Let us assume that in the sample space there is a set of points $\{\mathbf{A}_r\}$ ($r = 1, \dots, n$) and their number of $n \geq l$, where l is the dimension of the parameter vector $\boldsymbol{\theta}$. (As elements of the set there can be independent repeated measurements of harmonics of the same diffraction line $\{HKL\}$.) Then the problem of l- dimensional estimating of $\boldsymbol{\theta}$ can be reduced to l one-dimensional problems of parallel componentwise estimating of θ_q ($q = 1, \dots, l$) at different points of the sample space. Randomization of points is a way to limit of the estimates bias due sampling.

The measurements of $\mathbf{A}$ are considered as initial approximations to the true harmonics of $\hat{\mathbf{A}}$. Each measurements vector from the set $\{\mathbf{A}_r\}$ is given a random vector of auxiliary variables $\mathbf{u}$. Correlation of variables $(\mathbf{A}, \mathbf{u})$ can be neglected.

A solution realizing the asymptotically normal form of the maximum likelihood under constraints on the parameters $\boldsymbol{\theta}$ in the form of Eq. (7.4) will be the stationary point of the Lagrange function [2]:

$$\Lambda = \sum_{r=1}^{n} \left\{ \sum_{k=1}^{K} \left[\frac{1}{2} \left(\hat{A}_k - A_k \right) / \sigma_{A_k}^2 + \right. \right.$$

$$\left. \left. + \frac{1}{2} \left(\hat{u}_k - u_k \right)^2 / \sigma_{u_k}^2 + \lambda_k \, g_k \left(A_k, \boldsymbol{\theta} \right) \right] \right\}_r .$$

Here, $\sigma_{A_k}^2$ are the variances of the measurements A_k, whose estimates are available from the experiment, $\sigma_{u_k}^2$ are the variances of the variables u_k that estimates to be refined in the process of optimizing the objective function, and λ_k are the Lagrange multipliers.

At each step of the optimizing sequence in the parameter space:

$$\theta_q^{i+1} = \theta_q^i + \Delta\theta_q^i \left(\mathbf{A}_r, \boldsymbol{\theta}^i \right) \quad (i = 0, 1, \dots)$$

the measurements vector $\mathbf{A}_r$ for ongoing estimate θ_q is chosen randomly from the set $\{\mathbf{A}_r\}$. For approximate values of the remaining components of the vector $\boldsymbol{\theta}$ are taken their previous estimates on other independent measurements.

A starting point of the iterative process is any random vector $\boldsymbol{\theta}$ whose components θ_q ($q = 1, \dots, l$) are within the allowable intervals $\left[\underline{\theta_q}, \overline{\theta_q}\right]$ and are related to each other by the coefficient h_1^* in the adequate with data $\{\mathbf{A}_r\}$ regression Eq. (7.1).

In order not to complicate the objective function with the conditions for the optimization region the estimated parameters are represented by a bounded function of some variable $\boldsymbol{\omega}$ on an infinite interval:

$$\theta_q = \frac{1}{2}\left[\underline{\theta_q} + \overline{\theta_q}\right] + \frac{1}{2}\left[\overline{\theta_q} - \underline{\theta_q}\right]\sin\omega_q.$$

Derivatives of the functions of the model Eq. (7.4), which are involved in the optimization procedure, will be of the form

$$\partial g_k / \partial \omega_q = x_k H_q \cos\omega_q,$$

$$H_q = \frac{1}{2}\left[\overline{\theta_q} - \underline{\theta_q}\right]\prod_{s=1}^{4} e^{\theta_s}.$$

To calculate a step of the iterative optimization process there are obtained simple formulas:

$$\Delta\omega_q = \frac{\sum_{k=1}^{K} w_k e_k \left(\partial g_k / \partial \omega_q\right)}{\sum_{k=1}^{K} w_k \left(\partial g_k / \partial \omega_q\right)^2},$$

$$\lambda_k = w_k\left[\left(\partial g_k / \partial \omega_q\right)\Delta\omega_q - e_k\right].$$

These include the weighting coefficients w_k and the generalized deviations e_k calculated at the sample point r:

$$w_k = \left(\sigma_{A_k}^2 / \hat{A}_k^2 + \sigma_{u_k}^2\right)^{-1},$$

$$e_k = (\hat{A}_k - A_k)/\hat{A}_k + (\hat{u}_k - u_k) - g_k.$$

Required $\left(\hat{A}_k, \hat{u}_k\right)$, for which is true the model Eq. (7.4), initially are taken equal to measured values (A_k, u_k). Then using the Lagrange multipliers λ_k, there are calculated the running corrections

$$\Delta\hat{A}_k = -\left[(\hat{A}_k - A_k) + \hat{A}_k^{-1}\lambda_k\sigma_{A_k}^2\right],$$

$$\Delta\hat{u}_k = -\left[(\hat{u}_k - u_k) + \lambda_k\sigma_{u_k}^2\right].$$

Variances $\sigma_{u_k}^2$ are estimated at each iteration by the standard deviations of $\hat{u}_{k,r}$ ($r = 1, \ldots, n$) from the mean.

Iteration step adjustable by the coefficient $0 < \zeta^i \leq 1$ ($i = 0, 1, \ldots$):

$$\boldsymbol{\theta}^{i+1} = \boldsymbol{\theta}^i + \zeta^i \Delta\boldsymbol{\theta}^i(\Delta\boldsymbol{\omega}^i),$$

$$\hat{\mathbf{A}}^{i+1} = \hat{\mathbf{A}}^i + \zeta^i\Delta\hat{\mathbf{A}}^i,\ \ \hat{\mathbf{u}}^{i+1} = \hat{\mathbf{u}}^i + \zeta^i\Delta\hat{\mathbf{u}}^i$$

is considered to be allowable if the functional of weighted discrepancies is decreased, that is,

$$G^{i+1} < G^i = \sum_{r=1}^{n}\left\{\sum_{k=1}^{K} w_k^i \left|g_k^i\right|^2\right\}_r.$$

As a rule, a local minimum G is achieved in three or four iterations. If in that the deviations from the stationarity conditions of the objective function of Λ for all points of the sample space ($r = 1, \ldots, n$) are less than allowable errors:

$$\left|\sum_{k=1}^{K}\left(\partial g_k/\partial\omega_q\right)\lambda_k\right| < \varepsilon\ \ (q = 1, \ldots, l),$$

$$|\,\Delta\hat{A}_k\,| < \varepsilon,\ \ |\,\Delta\hat{u}_k\,| < \varepsilon\ \ (k = 1, \ldots, k_{\max}),$$

where $0 < \varepsilon \ll 1$, the optimum $\Lambda = \Lambda\left(\boldsymbol{\theta}^*\right)$ is likely being found. It is still necessary to verify that the obtained $\hat{\mathbf{A}}$ are within the confidence range of the measurements of **A** and that the vector of functions of the model $\mathbf{g}^* = \mathbf{g}\left(\hat{\mathbf{A}}, \boldsymbol{\theta}^*\right)$ no significantly deviates from zero.

For checking there are calculated statistical criteria

$$\begin{cases} \eta_1 = \dfrac{\zeta}{n}\sum_{r=1}^{n}\left\{\sum_{k=1}^{K}\left(\hat{A}_k - A_k\right)^2 / \sigma_{A_k}^2\right\}_r , \\ \eta_2 = \dfrac{\zeta}{n}\sum_{r=1}^{n}\left\{\sum_{k=1}^{K}\left|g_k^*\right|^2 / \sigma_{g_k}^2\right\}_r . \end{cases}$$

Correction $\zeta = 2/(1 - 4/nK)$ takes into account the presence in the model equation of two inexact variables and four estimated parameters. Unknown variances $\sigma_{g_k}^2$ can be estimated by the standard deviations of the residual discrepancies g_k^* from the means for n points.

For any distribution of $\{A_k\}$ and $\{g_k\}$ the solution is rejected with a reliability not smaller then P if either η_1 or η_2 is greater than $K/(1-P)$ [2].

Successful search results from random start points can be considered the independent realizations of a random vector $\boldsymbol{\omega}^* = \boldsymbol{\omega}(\boldsymbol{\theta}^*)$. In substance, by a randomized computational experiment, a random sample $\{\boldsymbol{\omega}_j^*\}$ ($j = 1, \dots, M$) is extracted from a completely unknown distribution, although it is assumed to be the same for all j. Obtained sample is used to constructing the confidence intervals for estimates of the model parameters of the object in given space

$$\mathbf{O} = \left[\rho_d,\ (\bar{\xi}/a),\ (\sigma_\xi/\bar{\xi}),\ (\bar{\tau}/b)\right].$$

The best for this purpose, there will be the self-correcting over sample confidence intervals that behave correctly under the transformations [15]:

$$\Re(P) = \bar{y} \pm \sigma\, t(P) + \sigma\, v\left[2\, t(P)^2 + 1\right] / 6\sqrt{M},$$

$$\sigma = \sqrt{\mu_2/(M-1)}, \quad v = \mu_3 / \mu_2^{3/2},$$

$$\mu_n = \sum_{j=1}^{M}(y_j - \bar{y})^n / M \quad (n = 2,\ 3).$$

Here, σ is the standard deviation of the sample mean $\bar{y}$; v is the asymmetry coefficient of sample; $t(P) = t_{(M-1),(1-\alpha/2)}$ are the percentage points of the t-distribution with $(M-1)$ degrees of freedom for a given confidence probability $P = 1 - \alpha$.

With a probability about of 99% no less than 90% of the realizations of any distribution of $y \equiv \overset{*}{\omega}_q$ $(q = 1, \ldots, l)$ are between the two extreme values of sample with the size $M = 60$ [37]. Thus, sample of size $M = 60$ can consider to be sufficient for constructing approximate confidence intervals such that confidence probability is not less than the given P.

The developed method is used to estimate the parameters of a random system of dislocations with large ($\bar{\xi}/a > 250$) or small ($\bar{\xi}/a < 50$) dislocation loops. With an intermediate level of loops sizes the estimation would to be carried out using two the model Eq. (6.17) with different Q_{HKL}, so there will require to introduce the matrix form of the method of Lagrange multipliers [2].

4. Verification of the method accuracy on model crystals with a given dislocation structure. For the optimal estimation of the parameters in different states of the dislocation structure of model crystals were selected the most informative diffraction lines (§ 6.3): {110} bcc when small loops of dislocations in iron crystals; {113} fcc when large loops of dislocations in aluminum crystals. The given observation interval in fractions of the reciprocal-lattice period is æ = 0.1. Parameters of the dislocation system in model crystals are listed in Table 7.1.

Table 7.1.

Identification of a system of dislocation loops by data of the simulation measurements

Model crystal	*Exact values of the parameters*	*Approximate 90% confidence intervals of estimates*	*Limitations from the preliminary regression analysis*
bcc (Fe) ⟨111⟩{110}	$\rho_d = 10^{11}$ cm^{-2} $\bar{\xi}/a = 25$ $\sigma_{\xi}/\bar{\xi} = 0.2$ $\tau/b = 2.0$ $\mu/\bar{\xi} = 0.2$	[1.0; 1.7]·10^{11} cm^{-2} [23; 27] [0.19; 0.22] [1.6; 1,9] [0.2; 0.3]	$\rho_d < (2.2 \pm 0.4) \cdot 10^{11}$ cm^{-2} $\bar{\xi}/a < 29 \pm 2$ $(k \geq 7)$
fcc (Al) ⟨110⟩{111}	$\rho_d = 10^{9}$ cm^{-2} $\bar{\xi}/a = 500$ $\sigma_{\xi}/\bar{\xi} = 0.2$ $\tau/b = 2.0^{*}$	[0.9; 1.2]·10^{9} cm^{-2} [460; 560] [0.19; 0.22] [1.5; 1,8]	$\rho_d \ll (2.3 \pm 0.5) \cdot 10^{10}$ cm^{-2} $(k \leq 20)$

* Taking into account the parameter of normal correlation ($\mu/d = 3.0$).

Measured values of the harmonics A_k were modeled by independent normally distributed random variables with theoretically calculated mathematical expectation $\hat{A}_k$ and with one and the same standard deviation $\sigma = 0.01$. Generated sample of vectors $\{\mathbf{A}_r\}$ $(r = 1, \ldots, n)$ is minimally necessary, that is $n = l = 4$. Deviations of data of the simulation measurements relative to the regression lines, by which the states of the structure are identified, are shown in Fig. 7.1.

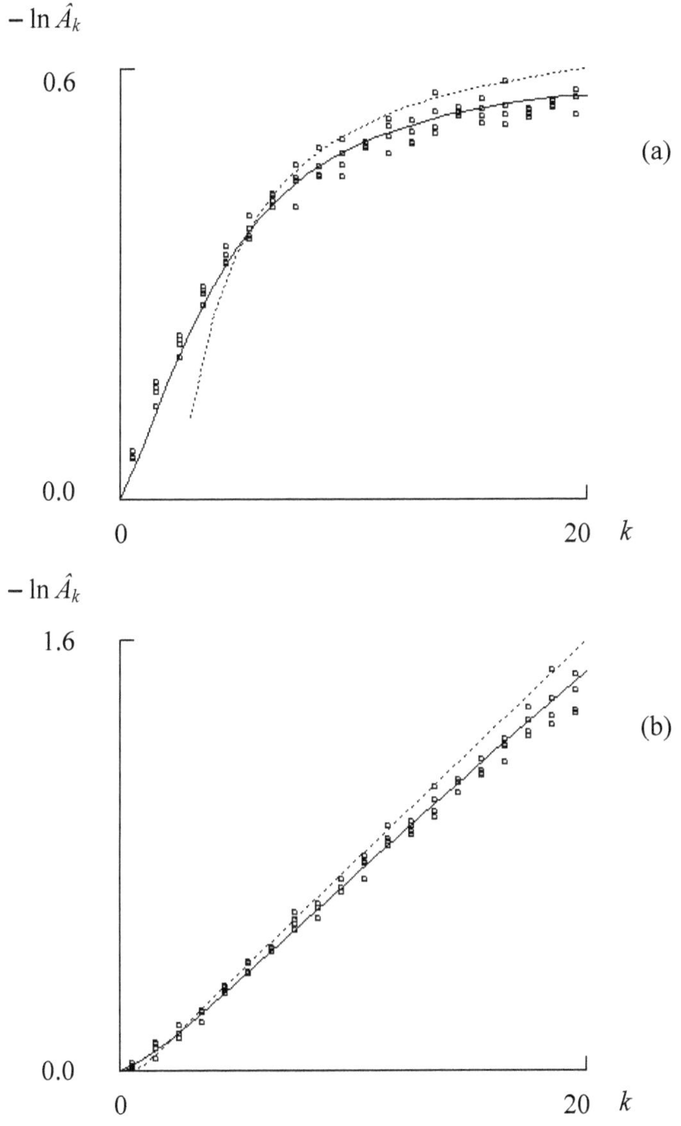

Fig. 7.1. Data of the simulation measurements for model crystals: (a) bcc – Fe; (b) fcc – Al. Strokes show deviations of approximate the model equations from calculated theoretical curves

According with regression estimating the level of dislocation loops sizes the limitations are established for local parametric identification:

$$\begin{cases} 10 < \bar{\xi}/a < 50 & \text{for small loops (Fe)} \\ 250 < \bar{\xi}/a < 1000 & \text{for large loops (Al)} \end{cases}$$

The smallest size of loops is compatible with the calculated value of the dislocation core that within $b \le r_0 \le 7b$ [55]. Approximate limiting larger size of loops is indicated by the electron microscopy of dislocation pile-ups.

Allowable values of the dislocation density follow from the experience; for variation coefficient of loops sizes and correlation radius there are stated the theoretically possible ranges (§ 6.2):

$$\begin{cases} 10^8 \le \rho_d \le 10^{12}\ \text{cm}^{-2}, \\ 0 \le \sigma_\xi/\bar{\xi} \le 0.3, \\ 0 \le \bar{\tau}/b \le 3.0. \end{cases}$$

Within the area of physical limitations all values of the model parameters of object are equally probable.

Results of verifying the method of the identification of system are summarized in Table 7.1. There are presented the average confidence intervals of estimates of structural parameters over 60 independent random samples with the size of 60 measured values of the vector $\boldsymbol{\theta}^*$.

The method developed correctly determined the structure of model crystals. Dislocation loops system is complete identifiable with the contribution of a priori information.

The sensitivity of estimates to the quality of a priori information has been studied in [73]. The distinguishability of sizes of large dislocation loops deteriorates with an increasing upper allowable boundary. What is important, this does not result in a serious loss of accuracy of the dislocation density estimates.

The effects of long-range order in the system of dislocations: If the ordering effects are significant, they are added to the systematic errors of the model of observations, and the identification is adapted to them.

When generating data for model crystals, the exact periodic component T_2 in the form of Eq. (6.13) and (6.16) was averaged over the ensemble of dislocation systems by the Monte Carlo method. As an approximate probability density of the parameter $1/\ell$ deviations about the expected value $\langle 1/\ell \rangle$, the Gauss function is taken with the standard deviation $\sigma_{1/\ell} / \langle 1/\ell \rangle = 0.1$.

Table 7.2.

Estimates of the main parameters of the dislocation system with availability of long-range order

Model crystal	*Specified parameter values*	*Approximate 90% confidence intervals of estimates*	
		Dislocations density ρ_d [cm^{-2}]	*Average radius of loops* $\bar{\xi}/a$
bcc (Fe) ⟨111⟩{112}	$\rho_d = 10^{11}$ cm^{-2} $\bar{\xi}/a = 25$ $\bar{\eta} = 0.75$ $\bar{\ell} = 3\bar{\xi}$	[0.7; 1.3]·10^{11} [0.9; 1.2]·10^{11}*	[22; 27] —
fcc (Al) ⟨110⟩{111}	$\rho_d = 10^{9}$ cm^{-2} $\bar{\xi}/a = 500$ $\bar{\eta} = 0.50$ $\bar{\ell} = 2\bar{\xi}$	[0.6; 0.8]·10^{9} [0.8; 1.0]·10^{9}*	[461; 556] —

* Method for estimating the density of dislocations when additional parameters limited in the allowable region (§ 7.4 and § 8.2).

Table 7.2 shows the obtained estimates of the main parameters of the system of dislocations when there are correlation, as in Table 7.1, and long-range order in the distribution of loops over the slip planes.

Analysis of harmonics of the diffraction line in different states of the structure gave estimates with insignificant bias relative to the true dislocation density.

Comparative estimates of the dislocation density by the width of the diffraction lines: By M.A. Krivoglaz theory the dislocation density ρ_d calculated from the width of the diffraction line at all does not depend on size of loops when these are considered a large. A network of large dislocation loops broadens the line like chaotically distributed rectilinear dislocations [39].

Equations by M.A. Krivoglaz for the diffraction lines of crystals with dislocation loops resulted in the following formulas:

$$\rho_d \approx \begin{cases} \dfrac{1}{2}\left(\dfrac{\gamma\Gamma}{a}\dfrac{æ}{Q_{HKL}}\dfrac{\sigma_D}{2\pi}\right)^2, & \text{very strong crystals distortions;} \\ \dfrac{24}{\bar{\xi}}\left(\dfrac{\gamma\Gamma^{-1}}{a}\dfrac{æ}{Q_{HKL}}\dfrac{B}{\pi}\right), & \text{not very strong crystals distortions.} \end{cases}$$

Here, σ_D is the variance of the theoretically Gaussian diffraction intensity distribution; B is the integral width of the distribution in Cauchy (or Lorenz) form. Both formulas are valid near the diffraction peak with a large Q_{HKL} only the limitations to first formula are stricter. Those are multipliers of $\gamma = a/b$, $\Gamma \sim 10$ [39].

For model crystals of iron and aluminum the central parts of the diffraction lines (æ = 0.025) were recovered using the exact theoretical harmonics $\{\hat{A}_k\}$. The value of σ_D is determined from the second moment of the diffraction maximum predicted by the harmonics. The integral width B is obtained by numerically integrating the function represented by the Fourier series. Thereto has checked stability of the results when increasing dimension of the harmonics vector $\{\hat{A}_k\}$. The values of ρ_d for different methods of calculation are given in Table 7.3.

Table 7.3.

Estimates of dislocation density by the width of the diffraction lines in different states of structure of model crystals

Diffraction line	*Model parameters*	*Estimated value*	
		$\rho_d(\sigma_D)$	$\rho_d(B, \bar{\xi}/a)$
{110} Fe	$\rho_d = 10^{11}$ cm^{-2} $\bar{\xi}/a = 25$	$2\cdot10^{11}$ $(5\cdot10^{11})^*$	10^{11} $(2\cdot10^{11})^*$
{113} Al	$\rho_d = 10^{9}$ cm^{-2} $\bar{\xi}/a = 500$	$3\cdot10^{10}$ $(5\cdot10^{10})^*$	$3\cdot10^{9}$ $(9\cdot10^{9})^*$

* When there is a correlation as in Table 7.1

If to raise the dislocation density in a model with large loops up to 10^{11}cm^{-2}, the formula of Gaussian broadening σ_D shows a value close to the true value of $\rho_d \sim 3\cdot10^{11}$ cm^{-2}. Distortion of model crystals of aluminum is not so strong in order to a peak acquired a Gaussian shape, and the estimate of dislocation density proved to be overstated by more than an order of magnitude. Formula of the integral width B in principle fine estimates the dislocation density when the dislocation loops are small and of precisely known size.

Usually it is used the formula for very strongly distorted crystals in assuming of rectilinear dislocations.

§ 7.2. Examination on Practical Identifiability of the Dislocation Structure for Test Specimens

For practical tests used the data of measuring the diffraction line {110} the 01Kh5 steel in martensite-quenched and 50%-deformed states that were received by D.A. Kozlov.

X-ray measuring has been made in the Fe K_α radiation under the greatest possible range of $\Delta 2\theta_{\{110\}} = 5.4$ degrees with step 0.1° at which the counted pulses uncorrelated. For randomization of instrumental errors were carried out four passes of line at point counting time of 10 s. Complete X-raying repeated after new install of specimens. The ratio of observation interval to the reciprocal lattice period of Fe crystals consists of æ = 0.12.

The reference line used as an instrumental profile was measuring on the annealed specimen. Method of optimal estimation of harmonics of the physical profile of the measured diffraction line is described in Ch. 9.

Sequential regression analysis of diffraction observations to assess the state of the structure of specimens goes through two stages.

In the first stage, a length of the interval [1, k] is revealed, where the approximation of initial harmonics by Eq. (7.3) becomes inadequate with the reliability $P > 0.5$ when the criterion $\eta > 2$.

$$\text{Quenched specimen: } \eta = \begin{cases} 3.0 & (k \le 4), \\ 6.3 & (k \le 5). \end{cases}$$

$$\text{Deformed specimen: } \eta = \begin{cases} 2.6 & (k \le 13), \\ 2.9 & (k \le 14). \end{cases}$$

Discrepancy between the measurement data of quenched specimen and the regression Eq. (7.3) is revealed at the very beginning of the observation interval, and it sharply increases. Deviations of the measurements of the deformed specimen from the approximating straight line of Eq. (7.3) increase gradually towards the end of the observations interval that is expectable since the real observations $\{Y_k\}$ contain systematic errors proportional to k^2, as consequence of splitting the diffraction doublet $K_{\alpha_1} - K_{\alpha_2}$ (§ 7.1).

At the second stage to account for systematic deviations from regression model in the basis functions vector $\mathbf{z}_k = (z_{k0}, z_{k1})$ it is enabled component $z_{k2} = (k/æ)^2$.

In Table 7.4 the estimates of regression equations are presented for stand out subsets of observations in their own bases over states of the structure.

Table 7.4.

Classification of the dislocation structure states of the test specimens

Structure forming process	*Coefficients of regression equations*			*Structure parameters limitations*
	h_0	h_1	h_2	
Martensite transfor-mation	0.0132 ± 0.0003	0.3326 ± 0.0155	0.000138 ± 0.000001	$\rho_d < (1.6 \pm 0.3) \cdot 10^{11}\,\text{cm}^{-2}$ $\bar{\xi}/a < 25 \pm 2$ $(5 \le k \le 17)$
Plastic deformation (50%)	0.000379 ± 0.000408	0.000529 ± 0.000023	0.000033 ± 0.000002	$\rho_d \ll (4.1 \pm 0.2) \cdot 10^{10}\,\text{cm}^{-2}$ $(1 \le k \le 17)$

The coefficient h_2 is significantly higher for the quenched specimen than for the deformed one, although herein the splitting of the martensite doublet is an order of magnitude smaller than the splitting of the doublet $K_{\alpha_1} - K_{\alpha_2}$. An additional systematic errors arise owing to the thin-plateably form of martensite crystals (§ 6.1).

Agreement with the measurement data within indicated ranges is not rejected by statistical checks: the probability that the model does not correspond to the data, $P < 0.5$. Dispersion of measurements and the regression curves for test specimens are shown in Fig. 7.2.

Sequential regression analysis of the data for 01Kh5 steel reliably reveals that the dislocation structures that arise during the martensitic transformation and during plastic deformation belong to different classes over the level of dislocation loops sizes. Conclusion does not disagree with physical representations. The deforming in rolling can really create large dislocation loops. The jump-like process of martensite transformation is accompanied by local plastic deformations for relieving stresses and the formation of large loops of dislocations is not anticipated.

When large dislocation loops in the structure of plastic deformation the estimate of the coefficient h_0 is completely uncertain (Table 7.4). To reveal the spatial correlation of large loops there are very large æ required for lines with large $\{HKL\}$.[1]

In the model for structure of bcc crystals two types of dislocation loops are assumed to be equiprobable: $\langle 111\rangle\{110\}$, $\langle 111\rangle\{112\}$. The limitations in the space of parameters of the model are the same as in § 7.1.

---

[1] M.A. Krivoglaz came to the conclusion that precisely at these conditions in principle there is measurable the density of dislocations when large network and correlation within a sphere for radius of “shielding” [40].

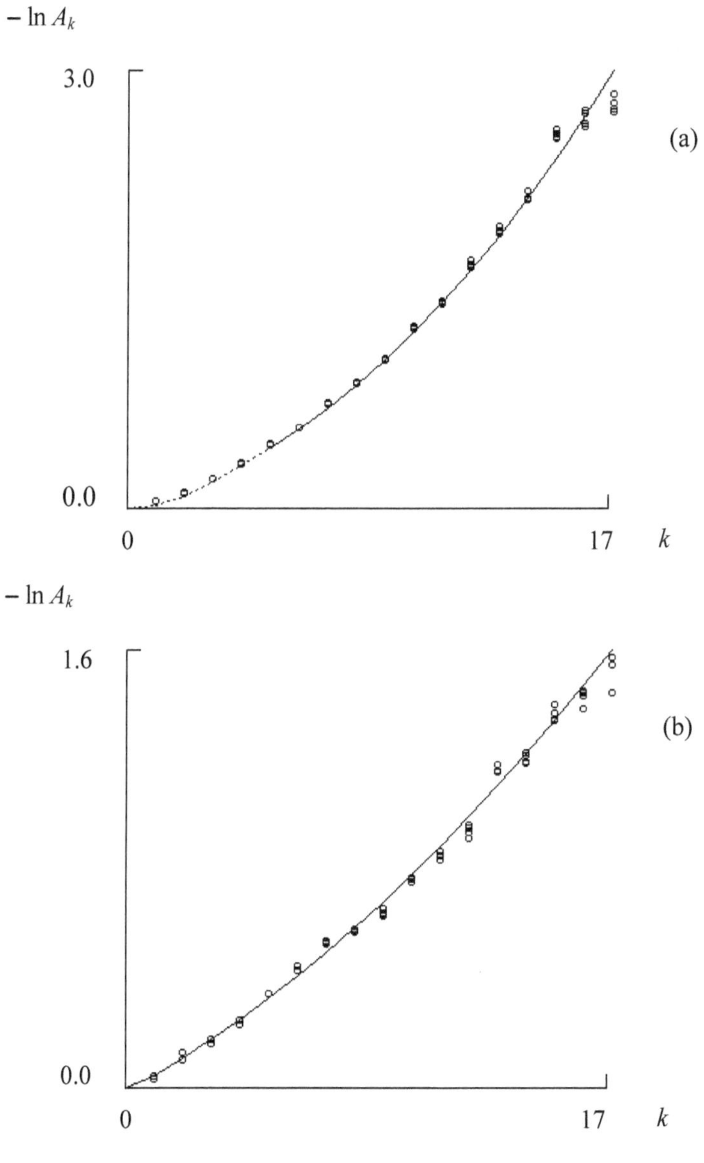

Fig. 7.2. Regression curves of the diffraction observations for 01Kh5 steel specimens with different processing:
(a) quenching for martensite; (b) rolling to strain of 50%

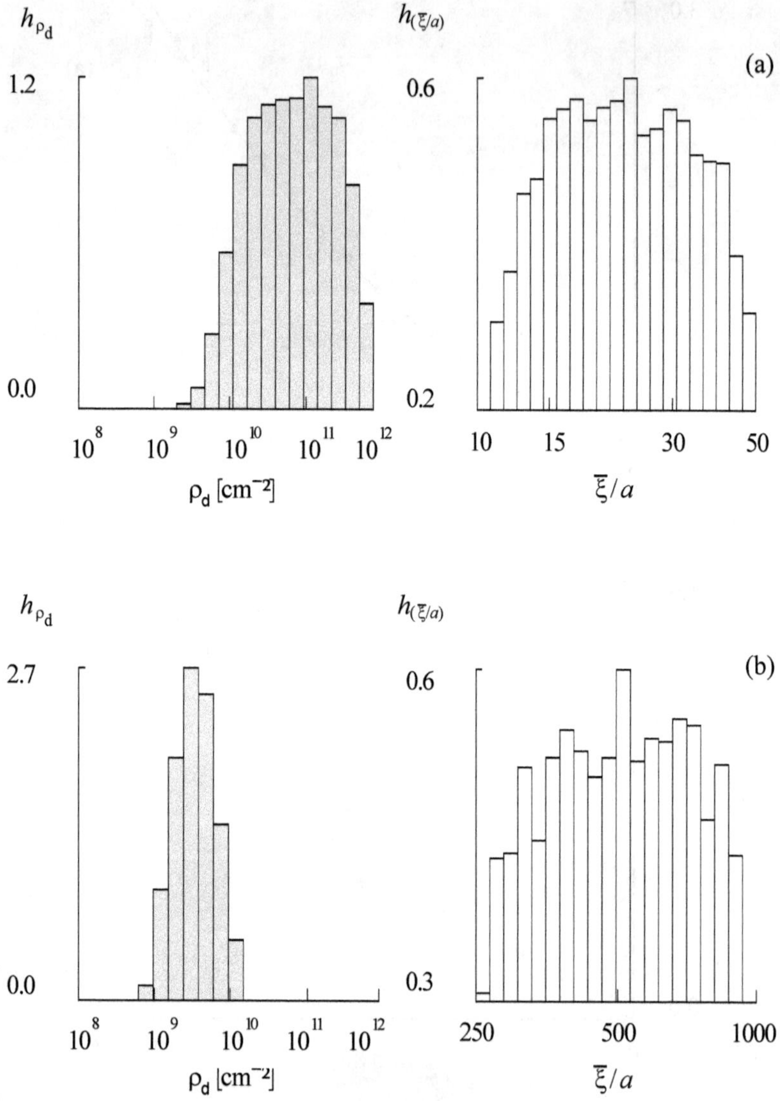

Fig. 7.3. Empirical distribution of the measurements of dislocation density (ρ_d) and average loops sizes ($\bar{\xi}/a$) depending on the type of structure: (a) martensite transformation; (b) plastic deformation of 50%

Search for an acceptable estimate of parameters with a random choice of an allowable starting point is considered as a computational experiment. Estimate is acceptable if an agreement of the model with the data of the diffraction observations is not rejected by statistical criteria. Acceptable estimate is considered to be a measured value of the vector of structural parameters.

For each test specimen in the course of computational experiments on their own model of diffraction observations were accumulated 9×10^3 of measurement results.

Figure 7.3 shows the distribution of measured values of dislocation densities and loops sizes. The frequency of occurrence of random variables in the intervals of the grouping is normalized to the width of the interval and sample size.

The histogram indicates approaching to the point estimate of the maximum likelihood, corresponding to the maximum of density of the a posteriori distribution when uniform a priori distribution of parameter in allowably region [2].

Table 7.5.

Estimates of the dislocation system parameters in the specimens of steel 01Kh5

Structure forming process	*Approximate 90% confidence intervals*			
	Dislocations density ρ_d [cm^{-2}]	*Average radius of loops* $\bar{\xi}/a$	*Variation coefficient* $\sigma_\xi/\bar{\xi}$	*Correlation radius* $\bar{\tau}/b$
Martensite transformation	$[0.7; 1.2]\cdot10^{11}$ $(1.0-1.6)\cdot10^{11*}$	[21; 26] (21– 22)*	[0.19; 0.22] —	[1.5; 1.9] (2.0 – 2.1)*
Plastic deformation (50%)	$[3.1; 4.0]\cdot10^{9}$ $(2.5-4.0)\cdot10^{9*}$	[456; 522] (500 – 536)*	[0.18; 0.22] —	[1.4; 1.8] (2.0 – 2.1)*

* The most plausible estimates by histograms maximum.

Presented in Table 7.5 confidence intervals for parameters of the system of dislocations in considered structures there are produced on the average 900 random samples of size 60 extracting from total sample of data. The nearly maximum-likelihood estimates of the main parameters are also given in Table. 7.5.

In practice, analysis of a random system of dislocations given rise to allowable, from a physical point of view, description of dislocation structure. The best accuracy of the parameters for large loops of dislocations in the structure of plastic deformation will be ensured when choosing the most informative diffraction line, precisely with a large value of Q_{HKL}.

With the quality level reached of the model of diffraction observations, a reliable identification of the dislocation structure of the deformed crystals is possible with minimum expenses to the experiment.

§ 7.3. Determination of Long-range Order in the Dislocation Structure of Crystals

The system of dislocation loops with periodic arrangement on slip planes models the structure of martensitic transformation (small loops), as well as slip bands in the structure of plastic deformation (large loops). With large loops of dislocations, the long-range order is practically inaccessible to observation. The order parameters in the system of small loops of dislocations are in principle measurable by the initial harmonics of the diffraction line with large indices $\{HKL\}$ (§ 6.3).

1. Diffraction observations model for estimating the ordered system of dislocations. The general equation of the harmonics of the diffraction line is transformed into a model of observations, which makes it possible to measure long-range order parameters simultaneously with the dislocation density:

$$\left.\begin{aligned}
&A_k \cong f_k(\boldsymbol{\Theta}) \quad (k < k_{\lim}),\\
&f_k(\boldsymbol{\Theta}) = \exp\left[-t e^{\varphi_0} - x_k e^{\varphi_1} - y_k e^{\theta_1}\left(\theta_2 e^{\theta_3}\right)\right],\\
&t = -C_0 Q_{HKL}^2,\\
&x_k = C_1 Q_{HKL}(k/æ), \quad y_k = \tfrac{1}{4} C_2 (k/æ)^2;
\end{aligned}\right\} \tag{7.5}$$

$$\boldsymbol{\Theta} = \begin{Bmatrix} \boldsymbol{\varphi} \\ \boldsymbol{\theta} \end{Bmatrix}, \quad \boldsymbol{\theta} = \begin{Bmatrix} \theta_1 = \ln\left[\bar{c}(\xi/a)\right] \\ \theta_2 = \bar{\eta} \\ \theta_3 = \ln\left[\overline{(\xi/\ell)^3 W}\right] \end{Bmatrix},$$

$$\boldsymbol{\varphi} = \begin{Bmatrix} \varphi_0 = \ln\left[\bar{c}(\xi/a)^2 J_\xi (\mu/a)\upsilon(1-\bar{\eta})\right] \\ \varphi_1 = \ln\left[\bar{c}(\xi/a)^2 J_\xi \left(1+\upsilon(1-\bar{\eta})\right)\right] \end{Bmatrix},$$

$$W = J_\xi^6 J_\ell, \quad \begin{Bmatrix} J_\xi = \left[1+(\sigma_\xi/\xi)^2\right], \\ J_\ell = \left[1+3\left(\sigma_{1/\ell}/(\overline{1/\ell})\right)^2\right]. \end{Bmatrix}$$

The area of definition of the observations model, within which the approximation of the diffraction theory equation is exact, is limited by the harmonics order of $k_{\lim}$.

When averaging the periodic component, the expected value of the function of the random variable $1/\ell$, which probability of deviation about the distribution center $\langle 1/\ell \rangle$ rapidly decreases (§ 6.3), is calculated using the expansion to the second central moment [105]:

$$\left\langle (1/\ell)^3 \right\rangle \approx \overline{(1/\ell)}^3 \left[1 + 3\left(\sigma_{1/\ell} / \overline{(1/\ell)} \right)^2 \right], \ \sigma_{1/\ell} / \overline{(1/\ell)} \le 0.1 .$$

In the model of diffraction observations, the vector of parameters $\boldsymbol{\Theta}$ has two components: $\boldsymbol{\theta}$ – the being determined parameters of the object; $\boldsymbol{\varphi}$ – other parameters of the dislocation system modeling the object, which make an auxiliary role in optimizing.

The initial approximations of the parameters $\boldsymbol{\varphi}$ are determined on the most accurate estimating the regression coefficients of the data $Y_k = -\ln A_k$:

$$Y_k \cong h_0 z_{k0} + h_1 z_{k1} \ \left(z_{k0} \equiv t, \ z_{k1} \equiv x_k \right).$$

Vector $\{Y_k\}$ (k = 1, 2) includes harmonics for which we can neglect the quadratic in k /æ term of the regression equation.

If the regression equation is not rejected by statistical test, given in § 7.1, the coefficients (h_0, h_1) are positive and statistically significant, then the observations model is applicable for estimating the parameters vector of $\boldsymbol{\Theta}$.

For crystals with small dislocation loops in the region of limitations of the diffraction observations model (7.5) there can be a minimum required number of harmonics, therefore, should accept $k_{\text{lim}} = 3$.

2. Method of measuring the density of dislocations and the degree of order by the slip planes. Different approximations of the theory differently distort the calculated initial harmonics A_k relative to true harmonics $\hat{A}_k$. For example, when $k \to 0$, the inaccuracy of equation of the diffraction observations model decreases, but the inaccuracy of the object model itself – with an approximate description of the displacement field created by dislocation loops in the crystal – increases.

It can be expected that in reality, on substantially limited interval ($1 \le k \le k_{\text{lim}}$), systematic errors of the predicted harmonics of A_k are close to uniform.

Assume that the squares of deviations from the true harmonics of $\hat{A}_k$, which are made up of data errors and model inaccuracies, are proportional to the variance of the measurements of σ_k^2 with an approximately constant unknown factor of ζ. Let's redefine the problem as the estimating by a data sampling $\{\mathbf{A}_\nu\}$ with the covariance matrix

$$\zeta \mathbf{V_A} = \zeta \begin{bmatrix} \sigma_1^2 & \cdots & 0 \\ \vdots & \ddots & \vdots \\ 0 & \cdots & \sigma_{k_{\lim}}^2 \end{bmatrix}.$$

With a normal distribution of the sample with a covariance matrix proportional to the unknown factor, the maximum likelihood method leads to the following objective function of model optimization [2]:

$$\left. \begin{aligned} L(\boldsymbol{\Theta}) &= \frac{nm}{2} \log \sum_{r=1}^{n} \sum_{k=1}^{m} \left| \varepsilon_{k,r}(\boldsymbol{\Theta}) \right|^2 / \sigma_k^2, \\ \varepsilon_{k,r}(\boldsymbol{\Theta}) &= A_{k,r} - f_k(\boldsymbol{\Theta}). \end{aligned} \right\} \tag{7.6}$$

Here, n is a sample size of the original data; $m = k_{\lim}$ is the dimension of the vector of observations $\{A_k\}$.

The matrix of the second derivatives of the objective function of $\mathbf{H}$ and the gradient vector of $\mathbf{g}$ are calculated together with the coefficient ζ, which adjusts the model in the process of approaching the optimum point $\boldsymbol{\Theta}^*$ [2]:

$$\boldsymbol{\Theta}^{i+1} = \boldsymbol{\Theta}^i - \mathbf{H}^{-1}\mathbf{g} \quad (i = 0, 1, \ldots),$$

$$\begin{cases} \mathbf{H} = 2 \sum_{r=1}^{n} \sum_{k=1}^{m} \left(\partial f_k / \partial \boldsymbol{\Theta} \right)^{\mathrm{t}} w_k \left(\partial f_k / \partial \boldsymbol{\Theta} \right), \\ \mathbf{g} = -2 \sum_{r=1}^{n} \sum_{k=1}^{m} \left(\partial f_k / \partial \boldsymbol{\Theta} \right)^{\mathrm{t}} w_k \, \varepsilon_{k,r}(\boldsymbol{\Theta}), \\ w_k(\zeta) = \dfrac{nm}{2\sigma_k^2} \left[\sum_{r=1}^{n} \sum_{k=1}^{m} \left| \varepsilon_{k,r}(\boldsymbol{\Theta}) \right|^2 / \sigma_k^2 \right]^{-1} \quad (1 \le k \le m). \end{cases}$$

A search of the optimum is carried out in the region of allowable values of the parameters of the object model that are uniquely related with the vector $\boldsymbol{\theta}$:

$$\begin{cases} 10^8 < \rho_{\mathrm{d}} = \bar{c}\left(\overline{\xi / a}\right)\left(2\pi / a^2\right) < 10^{12} \ \mathrm{cm}^{-2}, \\ 0 < \bar{\eta} < 1, \quad 0.1 < \left(\overline{\xi / \ell}\right) < 1, \\ 0 < \sigma_\xi / \bar{\xi} < 0.3, \quad 0 < \sigma_{1/\ell} / \left(\overline{1/\ell}\right) < 0.1. \end{cases}$$

This involves additional information from the theory that diffraction line is detectable, when the period of ordered arrangement of the loops is in the interval $\left(\xi < \bar{\ell} < 0.1\Xi\right)$ under the assumption a weakly inhomogeneous distortions field in a crystal of size Ξ (Ch. 6).

To remain within the allowable region, moving from a random starting point $\boldsymbol{\theta}_0$, let us introduce the transformation of parameters:

$$\theta_q = \frac{1}{2}\left[\underline{\theta_q} + \overline{\theta_q}\right] + \frac{1}{2}\left[\overline{\theta_q} - \underline{\theta_q}\right] \sin\omega_q \quad (q = 1, 2, 3).$$

Operating vector of parameters $\boldsymbol{\omega}$ can vary in infinite limits.

For the starting point $\boldsymbol{\varphi}_0$, we choose random variables from the intervals of deviations from the regression coefficients: $h_0 \pm \sigma_{h_0}$, $h_1 \pm \sigma_{h_1}$ (for any distribution law, the probability that random deviations do not go beyond the established limits, $P = 0.5$ [37]).

Having a good initial approximation, it is possible to optimize $\boldsymbol{\varphi}$ without restrictions: the objective function will begin to rise earlier than the step, regulated by the coefficient $0 < \gamma^{(i)} \le 1$ ($i = 0, 1, \ldots$), will reach the limit of the allowable region $e^{\varphi_0} > 0$, $e^{\varphi_1} > 0$.

In order to reduce the correlation of estimates $\{\Theta_q\}$, the optimization step for each q will be calculated from random subsamples of data with the exception of one vector of observations $\mathbf{A}_{j_q}$: when calculating the q-th component of the gradient vector $\mathbf{g}$, the summation is carried out over $r \ne j_q$, where j_q is a random integer from the interval $(1, n)$.

Upon reaching the optimum point $\boldsymbol{\Theta}^*$, it is necessary to validate the model for agreement with the observations data and test the statistical significance of the model parameters estimates.

If the model fits the data, the residual deviations are biased estimate of the data errors, what to be test by statistical criterion involving correction on bias:

$$\eta = \left(n - (l+1)/m\right)^{-1} \sum_{r=1}^{n} \sum_{k=1}^{m} \left|\varepsilon_{k,r}\left(\Theta^*\right)\right|^2 / \sigma_k^2,$$

Considering that in ($n \times m$) equations of the model there are ($l + 1$) fitting coefficients (l is the dimension of the estimated parameters vector $\boldsymbol{\Theta}$), the minimum required of data sample size is $n > (l+1)/m$.

The model will be rejected as not adequate to the data with a degree of certainty of at least P if $\eta > m/(1-P)$ [2].

With a good agreement of the model with observational data, the matrix $\mathbf{H}^{-1}(\boldsymbol{\Theta}^*)$ approaches the covariance matrix of the parameters $\mathbf{V}_{\boldsymbol{\Theta}}$ suitable for test the statistical significance and non-correlation of the estimates in the vector $\mathbf{\Theta}^*$. Acceptable estimates are considered as measured values of object model parameters.

Running computational experiments with a random choice of an allowable starting point $\mathbf{\Theta}_0$, we obtain a sufficient sample of measurements to construct confidence intervals for the dislocation density $\rho_d = e^{\theta_1} \cdot \frac{2\pi}{a^2}$ and a degree of order $\overline{\eta} = \theta_2$, as well as a lower confidence limit for period of order $\overline{(\xi/\ell)} < e^{\theta_3/3}$.

Approximate confidence intervals, self-corrected over a sample of measured values of the parameters, are constructed using the method described in § 7.1.

3. Verification of estimating the ordered system of dislocations to accuracy using simulation experiments. The diffraction line {112}, best on observability of the long-range order in bcc-crystals, was chosen into interval of æ = 0.2 from the theoretical period. Parameters of the model crystal are given in Table 7.6.

The measured harmonic values of the diffraction line of A_k are modeled by independent normally distributed random variables with a theoretically calculated expectation $\hat{A}_k$ and with the same standard deviation $\sigma = 0.01$. Generated data sample of the minimum required size of $n = 4$.

Figure 7.4 shows dispersion of simulated data and the error of the observations model where allowing for the long-range order in the system of dislocations. Systematic deviations of logarithms in Fig. 7.4 are more deviations of the approximate beginning harmonics from the exact theoretical $\hat{A}_k$, but a little.

In Table 7.6 are presented the results of measuring the parameters of model crystals using the initial harmonics of the diffraction line ($k_{\lim} = 3$). The average confidence intervals for the being determined parameters are given, constructed from 60 independent random samples of size 60, extracted from a total sample of 900 measured values of the vector $\mathbf{\Theta}^*$.

Table 7.6.

Parameter estimates for the long-range order in the system of dislocation loops according to simulation data

Model crystal	*Specified order parameters*	*Approximate 90% confidence intervals of estimates*	
		Dislocations density ρ_d [cm^{-2}]	*Degree of order* $\overline{\eta}$
bcc (Fe)	$\overline{\eta} = 0.2$; $\overline{\ell} = 3\overline{\xi}$	[1.5; 2.2]·10^{11}	[0.22; 0.36]
$\langle 111 \rangle \{112\}$ $\rho_d = 10^{11}$ cm^{-2}	$\overline{\eta} = 0.5$; $\overline{\ell} = 2\overline{\xi}$	[1.7; 2.6]·10^{11}	[0.45; 0.61]
$\overline{\xi}/a = 25$	$\overline{\eta} = 0.8$; $\overline{\ell} = \overline{\xi}$	[2.5; 4.1]·10^{11}	[0.71; 0.82]

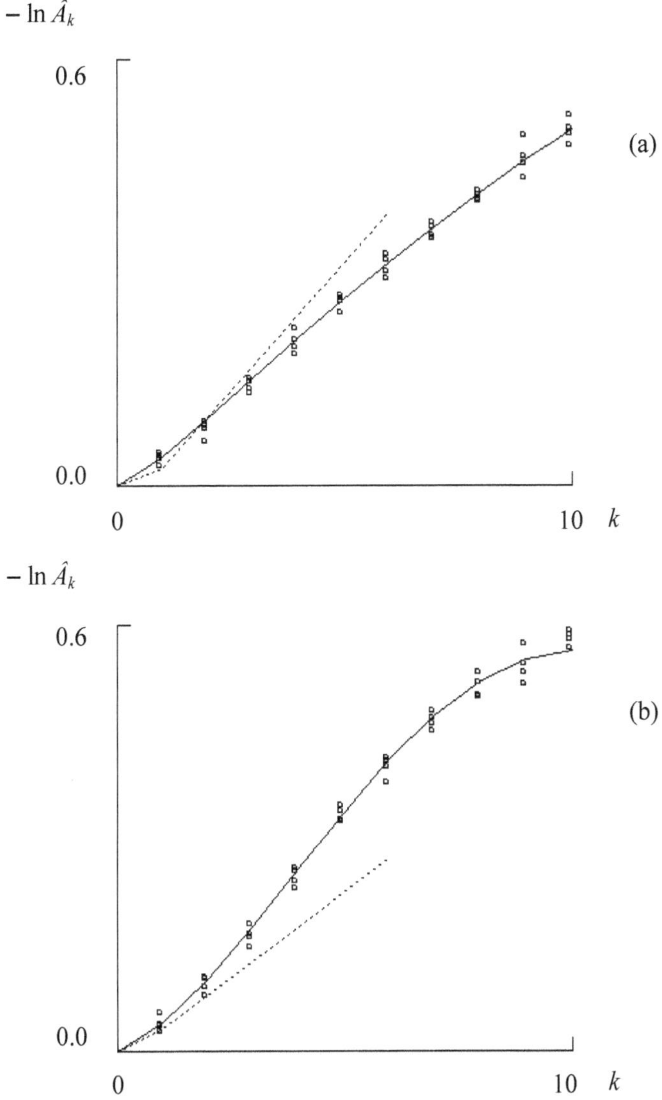

Fig. 7.4. Data of simulation measurements of the diffraction line of a model crystal with the long-range order in the system of dislocation loops: (a) $\overline{\eta} = 0.2$, $\overline{\ell} = 3\overline{\xi}$; (b) $\overline{\eta} = 0.8$, $\overline{\ell} = \overline{\xi}$. Solid line is exact theoretical equation and dashed line is approximate equation of the observations model

The order states in the dislocation structures of model crystals are distinguishable, despite of the estimates biases. The lower confidence limits of the long-range order period overlap when $\overline{\ell} > \overline{\xi}$.

The method yielded acceptable results of determining the long-range order in the system of small dislocation loops, even with a minimum sample of the original data.

§ 7.4. Method of Identification of the Dislocation Structure while Tetragonal Lattice of Crystals in Martensite of Steel

Object properties allow measuring the dislocations density by limiting other parameters of the dislocation structure model in a priori intervals [68].

1. Diffraction imaging of dislocations system in crystals of tetragonal symmetry. In the diffraction space, the bounded region of the observability of the system of dislocation loops, which results from the shearing martensite transformation, have been found:

$$\left.\begin{aligned} &A_k \cong e^{-\langle T\rangle_\xi} \quad (k \geq k_{\min}),\\ &T \cong 2\pi^2 \frac{v_0}{(2\pi)^3} q_i q_j \langle\Pi\rangle (\xi/a) \times\\ &\times \left[Z(\mathbf{R})(1+\upsilon) - \frac{1}{2\xi}\frac{\mu}{\xi}\upsilon \right] \frac{1}{p}\sum_{\alpha=1}^{p} \Gamma_{ij}^{(\alpha\beta)} \delta_{\alpha\beta},\\ &Z = \frac{4}{3\pi}\xi^{-1} - \frac{1}{4}|\mathbf{R}|^{-1} \quad (2\xi \ll |\mathbf{R}|); \end{aligned}\right\} \qquad (7.7)$$

$$|\mathbf{q}| = \frac{2\pi}{a} Q_{HKL}, \quad |\mathbf{R}| = \frac{k}{æ}\frac{a}{Q_{HKL}}, \quad Q_{HKL} = \sqrt{H^2 + K^2 + \left(\frac{a}{c}\right)^2 L^2}.$$

Harmonics of the diffraction intensity distribution A_k are determined by the function $T(\xi)$ averaged over loops sizes ξ, where $\mathbf{q}$ is the diffraction vector for crystallographic planes with indices $\{HKL\}$; (a, c) are the dimensions of the unit cell of a tetragonal crystal lattice of volume $v = a^2c$. The order of harmonics k corresponds to the observation range, whose fraction from the period in the reciprocal space is æ (§ 6.3).

The number of dislocation loops appearing in the crystal approximately is a Poisson random variable Π. The expected number of loops $\langle\Pi\rangle$ is large, but much smaller than the number of atoms in the crystal of N.

By parameter $\upsilon \sim \tau^2\mu$ there is measured the ellipsoid of correlation in the loops distribution when the correlations of their coordinates in the slip planes and over the normal thereto have the parameters $\tau, \mu \ll R$. The values of τ, μ are received equated to means for all p slip systems ($\alpha = 1, \dots, p$) assuming that variations of the correlation extent over slip systems can be neglected, since there are large fluctuations in the number of loops that fall into them when $\langle \Pi \rangle \ll N$.

Under scattering by crystals in the form of a thin plate, such as martensitic ones, the intensity distribution becomes asymmetric. Homogeneous deformations of crystals under transformation in a matrix shift the intensity distribution.

The exact Fourier coefficients of the distribution of the intensity of scattering by martensite crystals should to be written so (§ 6.1):

$$A_k + iB_k = e^{iq_j \varepsilon_{ij} R_i} \left\langle \left(e^{-\omega} + i\beta \right) e^{-T} \right\rangle_{\xi},$$

$$\beta(k) \ll 1, \quad \omega(k)/T(k) \ll 1.$$

Herein sine harmonics appear, and cosine harmonics decrease more quickly. Tensor of uniform extension of lattice ε_{ij} creates a shear coefficient.

In reality, since martensite crystals are strongly distorted by dislocations, the displacement and asymmetry of a significantly smeared distribution of the scattering intensity are became only by a small error in the model Eq. (7.7).

To calculate the diffraction on polycrystals, the sums of scattering intensities in the $\mathbf{q}_{HKL}$ direction are averaged by all possible random aggregates of crystals. The general diffraction equation for polycrystalline systems is constructed in § 8.1.

Let the Z axis of the Cartesian coordinate system (X, Y, Z) coincides with the normal to the reflecting plane $\{HKL\}$, and along it the diffraction vector $\mathbf{q}_{HKL}$ is directed. Fluctuations of the parameters of the system of dislocations in scattering crystals are neglected. Then the harmonics of the scattering intensity distribution along the Z axis are described by Eq. (7.7), where there is one nonzero component of the diffraction vector, precisely q_z, and of all the tensor coefficients only $\Gamma_{zz}^{(\alpha\alpha)}$ $(\alpha = 1, \dots, p)$ are retained. The mean $\Gamma_{zz}^{(\alpha\alpha)}$ for existing slip systems is transformed into the crystal-geometric coefficient of C_0.

When tetragonal body-centered lattice of crystals there are changed the Burgers vectors $\mathbf{b}$ and the normal vectors to slip planes $\mathbf{n}$:

$$\mathbf{b} = \frac{a}{2}\left[u\mathbf{i} + v\mathbf{j} + \frac{c}{a} w\mathbf{k} \right],$$

$$\mathbf{n} = \left[h\mathbf{i} + k\mathbf{j} + \frac{a}{c} l\mathbf{k} \right] \Big/ \sqrt{h^2 + k^2 + \left(\frac{a}{c}\right)^2 l^2},$$

(**i**, **j**, **k**) is an orthonormal basis, and $\langle uvw\rangle\{hkl\}$ are the indices of the slip systems.

In the numerical value of the crystal-geometric coefficient C_0 appears a correction ΔC_0 to the constant of bcc crystals (§ 6.3):

$$C_0 = 16.278295 + \Delta C_0,$$

$$\Delta C_0 = 12.4\,t + 7.75\,t^2 \quad \left(0 < t = \frac{c}{a} - 1 < 0.09\right).$$

Slip systems are $\langle 111\rangle\{110\}$, $\langle 111\rangle\{112\}$. The calculated interval is agreed with the experimentally determined maximum tetragonality of crystals in the martensite of steel [52].

An increase in the coefficient C_0 indicates that the tetragonal body-centered lattice undergoes a greater disordering by dislocation loops.[1]

2. Measurement of dislocation density in a tetragonal martensite with adaptation to systematic errors in the observations vector. Original data for identifying the dislocation structure formed during quenching of steel are the parameters of the model of the observed diffraction multipletlet. This is a narrow spectrum of lines that displays the phase state of the object.

In the model of diffraction spectrum the shape of all lines, the number of which n is considered known, is approximated by harmonics

$$A_k = \frac{\sum_{i=1}^{n} \hat{A}_{ki} w_i \cos(kz_i)}{\sum_{i=1}^{n} w_i \cos(kz_i)},$$

where $\hat{A}_{ki}$ are the true harmonics of spectrum lines, w_i and z_i are the weight fractions and shifts of lines relative to the origin.

In the theoretical interval of the Fourier representation, when the relative shifts of the lines are negligible, the harmonics of the approximate shape of spectrum lines are approximately equal to the weighted mean of the true harmonics.

The only $\{110\}$ multiplet that allows determination of parameters of the martensitic dislocation structure is contaminated with the $\{111\}$ line of fcc crystals of the matrix phase.[1]

---

[1] The fcc crystals are quite clearly more stable to the disordering of the periodic structure: $C_0 = 10.852197$ (§ 6.3).

Let us consider a theoretical example of the diffraction spectrum for a two-phase martensite with specified the phase composition (α–phases with a carbon concentration of 0.05 and 0.005 in fractions of 2:1) and the dislocation structure parameters $\left(\rho_d = 10^{12}\,\text{cm}^{-2},\ \bar{\xi}/a = 25\right)$. For the deformed state of the matrix γ–phase we can assume the allowable parameters $\left(\rho_d = 3\cdot 10^{9}\,\text{cm}^{-2},\ \bar{\xi}/a = 1000\right)$. Additional parameters of the dislocation system taken to be the same $\left(\sigma_{\xi}/\bar{\xi} = 0.2,\ \tau/b = 2,\ \mu/d = 2\right)$.

Figure 7.5 gives representation about distortions of harmonics of the true shape of the martensite line in observations data. An extent of distortion strongly depends on the fraction of the γ–phase with an alien dislocation structure, as in the model of diffraction spectrum the shape of all lines is averaged.

Under conditions of an increased uncertainty, it makes sense to put problem of determining the density of dislocations with the use of a priori information on the allowable values of the remaining parameters of the dislocation structure.

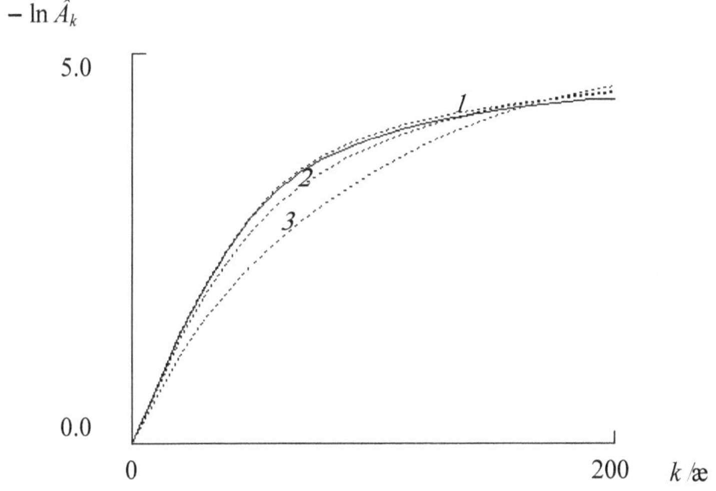

Fig. 7.5. Predicted distortions in the model of diffraction spectrum: solid line is the exact harmonics of the strong martensite component; dashed lines are harmonics of an averaged shape of the spectrum lines when the specified fraction of matrix γ–phase is (*1*) 0; (*2*) 0.05; (*3*) 0.20

---

[1] Precipitate of the smallest carbon particles does not affect the diffraction intensity distribution; only reduces it with regard to the diffuse background [39].

Equation (7.7) is transformed to the model of diffraction observations suitable for estimating parameters when existing distortions of the data:

$$\left.\begin{aligned} &A_k \cong f_k(\boldsymbol{\Theta}) + u_k, \\ &f_k(\boldsymbol{\Theta}) = \exp\left[-t\,e^{\theta+\varphi} - x_k\,e^{\theta+\psi}\right], \\ &t = C_0\,Q_{HKL}^2,\ x_k = -\frac{1}{2}C_0\,Q_{HKL}^3\,(k/æ)^{-1}\ (k \geq k_{\min}); \end{aligned}\right\} \qquad (7.8)$$

(t, x_k) are independent variables for the reflection $\{HKL\}$; u_k is an auxiliary variable that compensates the systematic errors of both the model and the data.

In the vector $\boldsymbol{\Theta} = (\theta, \varphi, \psi)^{\mathrm{t}}$ of estimated parameters as the major component is involved the average dislocations density:

$$\theta = \ln\bar{\rho}_{\mathrm{d}} + \ln\left[a^2/2\pi\right],\quad \bar{\rho}_{\mathrm{d}} = 2\pi\bar{\xi}\langle\Pi\rangle/Na^2c.$$

The remaining parameters of the dislocation system mixed after averaging $T(\xi)$ in Eq. (7.7) over the logarithmically normal distribution of loops sizes ξ.

Additional components of the vector N only indicate the upper boundary of the average radius of the dislocation loops:

$$\bar{\xi}/a < e^{\psi-\varphi}.$$

Mathematical expressions for the additional components φ and ψ are given in [68].

Data for estimating the parameters these are harmonics of the diffraction line A_k, that satisfy the restrictions of the observations model. The lowest order of the necessary harmonics is denoted $k_{\min}$.

Systematic biases of the higher order harmonics ($k \geq k_{\min}$) can be approximated by the orthonormal Legendre polynomials $P_l(z)$ [37].

Neglecting the differences in the shape of the spectrum lines, we obtain the sampling vectors of harmonics $\{\mathbf{A}_r\}$ ($r = 1, 2, \ldots, m$). The dimension of the vector $\mathbf{A} = \{A_k\}$ ($k = 1, 2, \ldots, K$) is equal to the highest order $k_{\max}$ of the well-defined harmonics of the physical profile of the measured multiplet.

Let us suppose that in sample all vectors $\mathbf{A}$ have the same distribution with the same covariance matrix

$$\mathbf{V}_{\mathbf{A}} = \begin{bmatrix} \sigma_1^2 & \cdots & 0 \\ \vdots & \ddots & \vdots \\ 0 & \cdots & \sigma_K^2 \end{bmatrix}.$$

The best estimate of the average dislocation density by available data of diffraction measurements is found at the minimum of the functional.

$$G(\mathbf{\Theta}) = L(\mathbf{\Theta}) + Z(\mathbf{\Theta}). \tag{7.9}$$

$$L(\mathbf{\Theta}) = \frac{1}{2}\sum_{k=k_{\min}}^{K}\sum_{r=1}^{m}\left|\varepsilon_{k,r} - \tilde{u}_k\right|^2 / \sigma_k^2,$$

$$\begin{cases} \varepsilon_{k,r} = A_{k,r} - f_k(\mathbf{\Theta}), \\ \tilde{u}_k = \dfrac{2}{M}\sum_{k'=k_{\min}}^{K}\bar{\varepsilon}_{k'}\sum_{l\geq 0}\dfrac{2l+1}{2}P_l(z_{j'})P_l(z_j), \\ z_j = (2j - M - 1)/(M-1),\ j = k - k_{\min} + 1, \\ M = K - k_{\min} + 1. \end{cases}$$

$$Z(\mathbf{\Theta}) = -\lambda\sum_{\nu=1}^{3}\left[\ln s_\nu + \ln(1 - s_\nu)\right],$$

$$\begin{cases} s_1 = (\theta - \underline{\theta})/(\overline{\theta} - \underline{\theta}), \\ s_2 = (\varphi - \underline{\varphi})/(\overline{\varphi} - \underline{\varphi}), \\ s_3 = (\psi - \underline{\psi})/(\overline{\psi} - \underline{\psi}). \end{cases}$$

As an assumption, there is taken the objective function of the maximum likelihood method of $L(\mathbf{\Theta})$ for sample with the nonuniform Gaussian distribution [105].

Approximation of the systematic biases u_k is carried out using the vector $\overline{\varepsilon} = \{\bar{\varepsilon}_k\}$ of the sample mean deviations $\varepsilon_{k,r}$ ($r = 1, 2, \ldots, m$) from the predicted harmonics $\hat{A}_k = f_k(\mathbf{\Theta})$ ($k_{\min} \leq k \leq k_{\max}$).

The penalty function $Z(\mathbf{\Theta})$ restricts the search for the optimum point $\mathbf{\Theta}^*$ in the allowable region of the parameter space. Vector $\mathbf{s}(\mathbf{\Theta})$ sets the upper and lower bounds of (θ, φ, ψ), following a priori intervals for the parameters of the dislocations system.

Coefficient λ in $Z(\mathbf{\Theta})$ reduces rapidly with approach to a minimum $G(\mathbf{\Theta})$. Suitable initial value it is $\lambda \sim 10^{-3}\, L(\mathbf{\Theta}^0)$, where $\mathbf{\Theta}^0$ is an allowable starting point. The minimizing sequence of points $\mathbf{\Theta}$ is calculated by the Marquardt-Gauss-Newton method [2].

At the minimum point $\mathbf{\Theta}^*$ it is required to calculate the criterion for checking the agreement of the diffraction observations model with data:

$$\eta = \frac{1}{m - \nu/M} \sum_{k=k_{\min}}^{K} \sum_{r=1}^{m} \left| \varepsilon_{k,r}^{*} - \tilde{u}_k^{*} \right|^2 / \sigma_k^2 ,$$

$$\varepsilon_{k,r}^{*} = \varepsilon_{k,r}\left(\mathbf{\Theta}^{*}\right), \quad \tilde{u}_k^{*} = \tilde{u}_k\left(\overline{\varepsilon}^{*}\right).$$

When u_k is approximated by Legendre polynomials of degree $l \leq l_{\max}$, the total number of fitting parameters in the M equations of the model (7.8) is $\nu = 3 + (l_{\max} + 1)$. It is taken into account in the correction for bias of the residual sum of squares.

The solution of $\mathbf{\Theta}^{*}$ is always rejected as not corresponding to the observations data with probability P, if $\eta > M/(1-P)$ [2].

For a lower sensitivity of the estimates of $\mathbf{\Theta}^{*}$ to the choice of the confidence probability P in practice it is realized the principle of stable minimization of functionals [95].

Successful optimization, proceeding from random points uniformly distributed in the allowable region of the parameter space of the object, will give a numerous of independent measurements of the average dislocation density in martensite crystals.

3. Test of the method for measuring the dislocation density in the martensite transformation structure. Reliability of the deriving information was verified when measuring the density of dislocations in a single-phase martensite of 01Kh5 steel with a lattice of crystals practically indistinguishable from cubic.

The results of two different methods for identifying the dislocation structure from the same vector of the most reliable harmonics of the physical profile of the diffraction line {110} ($k_{\max} = 15$) are compared in Table 7.7.

Table 7.7.

Dislocation density in martensite with cubic crystal lattice when different method of estimation of the model

Value to be estimated by the model of dislocation structure	*Approximate 90% confidence intervals of estimates*	*Maximum-likelihood estimates*
The density of dislocations with the limitation of "over" parameters	$[0.7; 0.8]\cdot 10^{11}$ cm^{-2}	$(0.6–1.0)\cdot 10^{11}$ cm^{-2}
The vector of parameters of the system of dislocations loops	$[0.4; 0.7]\cdot 10^{11}$ cm^{-2}	$(0.6–1.0)\cdot 10^{11}$ cm^{-2}

Maximum of the sampling distribution of the dislocation density measurements corresponds to the maximum-likelihood estimate (sample size of 9×10^3).

For homogeneous martensite with cubic symmetry of crystals the estimates of the density of dislocations by different methods differ only in accuracy, which seeks to prove the reliability of the results obtained.

4. The degree of order in the dislocation structure of tetragonal martensite crystals. Under § 7.3 there is described the method applicable for identifying long-range order in the dislocation structure of the martensitic transformation by the initial harmonics of the diffraction line with large indices $\{HKL\}$.

When tetragonal lattice of crystals, it is required to transform the independent variables of the diffraction observations model in the form of Eq. (7.5):

$$\left.\begin{aligned} t &= -\frac{c}{a}\left[C_0 + \Delta C_0\right] Q_{HKL}^2, \\ x_k &= \frac{c}{a}\left[C_1 + \Delta C_1\right] Q_{HKL}\left(k/æ\right), \\ y_k &= \tfrac{1}{4}\frac{c}{a}\left[C_2 + \Delta C_2\right]\left(k/æ\right)^2; \end{aligned}\right\} \quad \begin{aligned} Q_{HKL} &= \sqrt{H^2 + K^2 + \left(\frac{a}{c}\right)^2 L^2}, \\ æ &= æ\left(\frac{a}{c}\right) \ (\S\ 6.3). \end{aligned}$$

Here, to crystal-geometric coefficients, the numerical values of which are given in § 6.3, corrections were introduced depending on the lattice tetragonality degree $\frac{c}{a}$:

$\langle 111\rangle\{110\}$ bcc:	$\Delta C_1 = 1.26\,t,$	$\Delta C_2 = 5.20\,t;$
$\langle 111\rangle\{112\}$ bcc:	$\Delta C_1 = 2.12\,t,$	$\Delta C_2 = 5.20\,t.$

Estimated interval by $t = \frac{c}{a} - 1$ is the same as for the correction ΔC_0, given above (§ 7.4).

Providing for the tetragonality, the dislocation density to be measured of $\rho_d = e^{\theta_1} \cdot \frac{2\pi}{a^2}\left(\frac{a}{c}\right)$, where θ_1 is the component of the of parameters vector $\boldsymbol{\Theta}$ in the observations model Eq. (7.5).

Practical analysis of the dislocation structure of tetragonal martensite by two methods for diffraction multiplets $\{110\}$ and $\{112\}$ is presented on the example of a carbon steel sample in § 10.1.

CHAPTER 8
INHOMOGENEOUS DENSITY OF DISLOCATIONS IN A POLYCRYSTALLINE SYSTEM

Dispersion of orientations of the deforming crystals generates a microinhomogeneous dislocation structure. Occurrence of the orientational order (thence crystallographic texture) is accompanied by converting of inhomogeneity into a macroscopic one.

In the model of a complex dislocation structure of a polycrystalline system, the measuring parameters are fluctuations in the density of dislocations over crystals and structural components. Parameters describing the sizes and arrangement of the dislocation loops within crystals are only limited in the region of allowable values [74–75].

§ 8.1. Diffraction Imaging Structures of Plastic Deformation of Polycrystals

There are revealed the effects of inhomogeneous dislocation density in a polycrystalline system with the crystal orientation ordering inherent in a plane strained state.

1. Diffraction on polycrystals with a rhombic texture. Distribution of scattering intensity by a deformed crystal, when the effect of its sizes is negligible, is determined using Fourier image $I(\mathbf{R})$ (6.3):

$$J(\Delta\mathbf{q}) = NF^2 \sum_{\mathbf{R}} e^{-T(\mathbf{R})} e^{i\Delta\mathbf{q}\mathbf{R}}.$$

Here, $\Delta\mathbf{q}$ is a deviation from given diffraction vector of an ideal crystal $\mathbf{q}_{HKL}$.

If the system of Cartesian coordinates (X, Y, Z) is chosen so that the Z axis coincides with the normal to the reflecting plane $\{HKL\}$, then

$$\mathbf{q}_{HKL} = \left(0,\ 0,\ 2\pi/d_{HKL}\right),\ d_{HKL} = a\Big/\sqrt{H^2 + K^2 + L^2},$$

and a is period of the crystal lattice.

The intensity of scattering in the $\mathbf{q}_{HKL}$ direction by a crystal rotating around the Z axis can be found by integrating $J(\Delta\mathbf{q})$ over the cross sections of the reciprocal lattice site perpendicular to $\mathbf{q}_{HKL}$ within the unit cell of the reciprocal space. For the volume unit of the reciprocal space it is obtained the intensity

$$J(\Delta q_z) = NF^2 \left(d_z/2\pi\right) \sum_{R_Z} e^{-T(R_Z)} e^{i\Delta q_Z R_Z},$$

$$d_z \equiv d_{HKL},\ R_z = md_z\ \ (m = 0, 1, \ldots).$$

Since $-\pi / d_z \leq \Delta q_z \leq \pi / d_z$, setting $\Delta q_z = z / d_z$, where $-\pi \leq z \leq \pi$, we arrive at the Fourier representation of the distribution of the scattering intensity along the Z axis:

$$J(z) = NF^2 (2\pi)^{-1} \sum_m e^{-T(m)} e^{imz}.$$

In the diffraction space of a polycrystal, the intensities $J_j(z)$ ($j = 1, \ldots, r$) from the scattering in the direction $\mathbf{q}_{HKL}$ of a random number r of crystals with unequal distortions are summed. A random sum of intensities, averaged over all possible aggregates of scattering crystals, will show the diffraction line of a polycrystal:

$$J_D(z) = \left\langle \sum_{j=1}^{r} J_j(z) \right\rangle.$$

Profile of the line $J_D(z)$ is determined by the shape function of the reciprocal lattice site

$$G = \left\langle \sum_{j=1}^{r} e^{-T_j(m)} \right\rangle,$$

which depends on the distortions of each crystal in a scattering aggregate.

Let us represent the number of defects in an ensemble of deformed crystals by a multidimensional random variable $\mathbf{\Pi} \equiv \{\Pi_j\}$ ($j = 1, 2, \ldots$). A set of $\{\Pi_j\}$ is the realization of random increments of the number of defects in deforming crystals randomly oriented in the initial state of an ensemble. With a small error, we can assume that the random variables $\{\Pi_j\}$ are mutually independent. Then the multidimensional probability distribution of the vector $\mathbf{\Pi}$ is approximately the product of the Poisson distributions of its independent components $\{\Pi_j\}$.

Function $e^{-T_j(m)} \cong e^{-W\Pi_j}$ represents smearing a lattice site in the reciprocal space when there are $\Pi_j = N\overline{c_j}$ dislocation loops in the j-th crystal. Coefficient W is the measure of crystal distortion by dislocation loops with random sizes and correlated coordinates. Fluctuations of W on crystals will be neglected.

Let us consider a polycrystalline system consisting of s subsystems (texture components). We suppose first that for all crystals of one subsystem ($\nu = 1, \ldots, s$) independent random variables Π_j have the same Poisson distribution with the mathematical expectation $\langle \Pi \rangle_\nu$. If $W \ll 1$, the mathematical expectation of the random function $e^{-W\Pi_j}$ in the subsystem ν is following

$$\left\langle e^{-W\Pi_j} \right\rangle_\nu \cong e^{-W\langle \Pi \rangle_\nu}.$$

So, to describe a form of diffraction line of a polycrystal of $J_D(z)$ it is required to calculate the expected value of the sum of a random number of independent random variables $e^{-W\Pi_j}$ ($j = 1, \ldots, r$) identically distributed in a random subset $r_\nu < r$ with mathematical expectation $e^{-W\langle\Pi\rangle_\nu}$ ($\nu = 1, \ldots, s$).

Let the vectors of the cubic lattice $\langle pqr\rangle$ in the ν-th component of the rhombic texture of a polycrystal be ordered in the direction of the normal to the plane of a specimen that coinciding with the Z axis. Random deviations of the vectors $\langle pqr\rangle$ from the most probable direction can theoretically be distributed according to the normal law:

$$f_{\langle pqr\rangle} \sim \left\langle e^{K_\nu \cos\vartheta} \right\rangle,$$

where $K_\nu > 0$ is distribution parameter, and ϑ is deviation angle; the averaging is carried out over all the vectors $\langle pqr\rangle$ appearing under symmetry transformations of cubic crystals [71].

For some vector of type $\langle HKL\rangle$ rotated relative to the vectors $\langle pqr\rangle$ by an angle ϑ_ρ, the probability density of orientations will be

$$f_{\langle HKL\rangle} \sim \left\langle e^{K_\nu \cos(\vartheta - \vartheta_\rho)} \right\rangle,$$

and the probability of being within $\delta\vartheta \ll 1$ from the Z axis it is

$$\phi = \int_0^{\delta\vartheta} f_{\langle HKL\rangle} \sin\vartheta\, d\vartheta \cong \lambda \frac{K_\nu}{2\,\mathrm{sh}\,K_\nu} \left\langle e^{K_\nu \cos\vartheta_\rho} \right\rangle,$$

$$\lambda = 1 - \cos\delta\vartheta \approx \frac{1}{2}\left|\delta\vartheta\right|^2.$$

The probability that any of the vectors $\langle HKL\rangle$ is oriented close to Z is the sum of the probabilities ϕ.

Therefore, with the texture of type $\{pqr\}$ the scattering probability in the $\mathbf{q}_{HKL}$ direction is calculated by the formula

$$\Gamma_\nu \cong \lambda \frac{K_\nu}{1 - e^{-2K_\nu}} \sum_{\{HKL\}} \left\langle e^{-K_\nu(1-\eta)} \right\rangle_{\{pqr\}},$$

$$\eta = \frac{\left|Hp + Kq + Lr\right|}{\sqrt{H^2 + K^2 + L^2}\sqrt{p^2 + q^2 + r^2}}.$$

For a mixed distribution of the orientations of crystals, which is inherent in a multi-component texture, the scattering probability will be equal to

$$\Gamma = \sum_{\nu=1}^{s} \mu_\nu \Gamma_\nu .$$

Here, μ_ν are the weight fractions of texture components whose number is $s\left(\sum_{\nu=1}^{s} \mu_\nu = 1\right)$.

In the absence of texture, when $K_\nu \to 0 \ (\nu = 1, \ldots, s)$, it is obtained $\Gamma \to \frac{\lambda}{2} p_{\{HKL\}}$, where $p_{\{HKL\}}$ is the factor of repeated planes of the $\{HKL\}$ type.

Into a random sample among the general population out of the ν-th texture component can occur a random number of scattering crystals, which is calculated as the sum of mutually independent Bernoulli random variables [37]:

$$r_\nu = \sum_{i=1}^{M} \beta_i; \quad \beta_i = \begin{cases} 1 & \text{with probability } \Gamma_\nu, \\ 0 & \text{with probability } (1 - \Gamma_\nu); \end{cases}$$

M is an independent random variable with a Poisson distribution.

Using the generating functions of distributions, we find according to [37] that r_ν has a complex Poisson distribution with mathematical expectation $\langle r_\nu \rangle = \langle M \rangle \Gamma_\nu$ (and the same variance $\sigma^2_{r_\nu}$), where $\langle M \rangle$ is the average number of crystals falling in the specimen volume illuminated by a beam of rays.

All the expected number of scattering crystals is

$$\langle r \rangle = \sum_{\nu=1}^{s} \mu_\nu \langle r_\nu \rangle = \langle M \rangle \Gamma .$$

Note, by the way, that the fluctuations in the scattering capacity for a multicomponent texture increase and do not disappear for arbitrarily large $\langle M \rangle$:

$$\frac{\sigma_r^2}{\langle r \rangle^2} = \frac{1}{\Gamma^2} \left[\frac{\Gamma}{\langle M \rangle} + \sum_{\nu=1}^{s} \mu_\nu (\Gamma_\nu - \Gamma)^2 \right].$$

Here, it is used the formula of variance for a mixed distribution [31].

Now it can be found the mathematical expectation of a random sum in the function G, which describes the shape of the scattering intensity maxima in the reciprocal space of a polycrystal. And the equation of diffraction line takes the form

$$J_D(z) \cong NF^2 \langle M \rangle (2\pi)^{-1} \sum_m \left[\sum_{\nu=1}^{s} \mu_\nu \Gamma_\nu e^{-W\langle \Pi \rangle_\nu} \right] e^{imz}.$$

The entire intensity of scattering by a polycrystal per unit volume of the reciprocal space is

$$\overline{J_D} = \int_{-\pi}^{\pi} J_D(z)dz = NF^2 \langle M \rangle \sum_{\nu=1}^{s} \mu_\nu \Gamma_\nu.$$

Moving to the normalized intensities $J_D(z)/\overline{J_D}$, we make out that the diffraction line of a polycrystal in the interval, which is a fraction of æ from the period of the reciprocal lattice $2\pi/d_{HKL}$, has a representation in the form of a Fourier series with harmonics of order $k = æm$:

$$\left. \begin{aligned} A_k &= \sum_{\nu=1}^{s} \gamma_\nu A_k^\nu, \\ A_k^\nu &\cong e^{-W\langle \Pi \rangle_\nu}, \quad \gamma_\nu = \mu_\nu \Gamma_\nu / \Gamma. \end{aligned} \right\} \tag{8.1}$$

Thus, a multicomponent crystallographic texture gives rise to a diffraction multiplet with a different shape of lines entering it, represented by the harmonics $\{A_k^\nu\}$, and with weight γ_ν ($\nu = 1, \ldots, s$) varying depending on the type of reflection $\{HKL\}$ (and that is predictable for physical reasons).

According to the condition for the resolvability of the diffraction multiplet under the variances of the harmonics σ_ν^2 caused by the Poisson fluctuations of the number of defects in crystals:

$$\left| A_k^\nu - A_k^{\nu'} \right|^2 > \sigma_\nu^2 + \sigma_{\nu'}^2, \ \sigma_\nu^2 = \left(A_k^\nu W \right)^2 \langle \Pi \rangle_\nu \ (\nu' \neq \nu),$$

there is in principle indistinguishable difference of the average dislocation densities

$$|\Delta\rho_d| = \left| \rho_d^\nu - \rho_d^{\nu'} \right|, \ \rho_d^\nu = \langle \Pi \rangle_\nu (2\pi\xi)/V.$$

For example, when the average radius of loops $\bar{\xi} \sim 500\,a$, the average crystal size $\Xi \sim V^{1/3} \sim 10$ μm, it is indistinguishable $|\Delta\rho_d| \leq 10^7$ cm^{-2}.

Relative shifts of lines (ν', ν) in the multiplet, which are revealed when finite dimensions of crystals expanding from the dislocations, are of evaluation $|\Delta z_{\nu,\nu'}|/2\pi < |\Delta\rho_d| b^2/æ$ [63, 94]. For $æ \sim 0.1$ the shifts are negligible when $|\Delta\rho_d| \leq 10^{11}$ cm^{-2}.

Under the assumptions made about character of the distribution of random variables Π_j ($j = 1, 2, \ldots$) the expected total number of defects in all scattering crystals is

$$\left\langle \sum_{j=1}^{r} \Pi_j \right\rangle = \langle M \rangle \sum_{\nu=1}^{s} \mu_\nu \Gamma_\nu \langle \Pi \rangle_\nu .$$

The average number of defects that falls at one scattering crystal, the expected number of which is $\langle M \rangle \Gamma$, will be

$$\overline{\langle \Pi \rangle} = \sum_{\nu=1}^{s} \gamma_\nu \langle \Pi \rangle_\nu .$$

If assign to all crystals of a scattering aggregate the same distribution of the number of defects with average of a mixed Poisson $\overline{\langle \Pi \rangle}$, then there is received the equation of the harmonics of the diffraction line in the approximation of homogeneous structure of a polycrystal:

$$A_k \cong e^{-W\overline{\langle \Pi \rangle}} .$$

Equation (8.1) gets exactly the same form when $k \to 0$. This means that the initial harmonics will reveal the averaged structure.

For dislocation structure of plastic deformation applicable the following approximate expression

$$\left. \begin{aligned} W \cong \frac{1}{N} \left(\bar{\xi}/a\right)^2 \left[1 + \left(\sigma_\xi/\bar{\xi}\right)^2\right] \left[1 + (2\pi/p)\left(\bar{\tau}/b\right)^2\right] \times \\ \times \left(C_1\, Q_{HKL}\, m\right), \quad m/Q_{HKL} < 2\bar{\xi}/a. \end{aligned} \right\} \quad (8.2)$$

Since $C_1 < 10$ and $\bar{\tau}/b \ll \bar{\xi}/a$, the condition $W(m) \ll 1$ is valid for all allowable m when $\bar{\xi} < 0.1\Xi$ (§ 6.3).

Figure 8.1 gives a represent of the changing diffraction pattern when a macroscopic inhomogeneity of the dislocation density in a polycrystal.

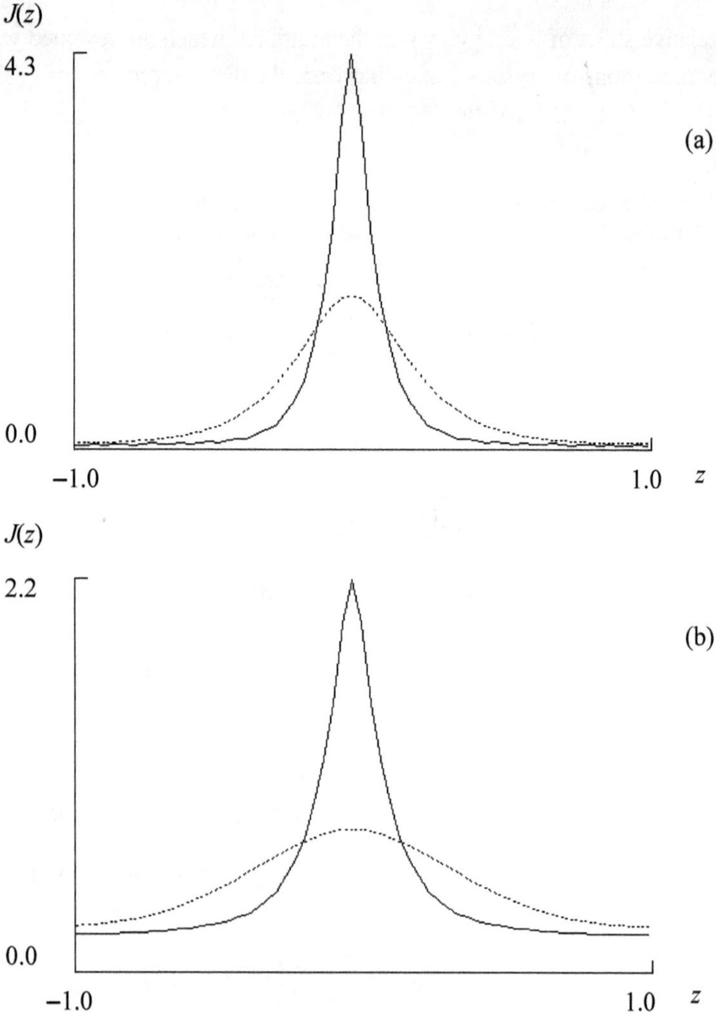

Fig. 8.1. The form of diffraction intensity distribution when macroscopic fluctuations of the dislocation density in scattering crystals: (a) $\gamma_2 = 0.1$; (b) $\gamma_2 = 0.5$. Dashed profile for a homogeneous aggregate of crystals with the same as number of defects

As an example is taken the diffraction line {112} bcc in the interval æ = 0.1 from the period of the reciprocal lattice. In numerical analysis was used the exact theoretical expression for e^{-T} (§ 6.2).

There are calculated $\hat{A}_k = \left[\gamma_1 \hat{A}_k\left(\rho_d^{(1)}\right) + \gamma_2 \hat{A}_k\left(\rho_d^{(2)}\right)\right]$, and $\hat{A}_k\left(\rho_d^{\Sigma}\right)$, where $\rho_d^{\Sigma} = \gamma_1 \rho_d^{(1)} + \gamma_2 \rho_d^{(2)}$ for given $\left(\rho_d^{(1)} = 10^9\ \text{cm}^{-2},\ \rho_d^{(2)} = 10^{11}\ \text{cm}^{-2}\right)$ and varying ratios of the weight fractions of γ_2/γ_1 in an aggregate of scattering crystals. The remaining structural parameters are accepted on average as in model crystals with large dislocation loops (§ 7.1).

By uniform distribution of the total number of defects in scattering crystals as compared to the mixed distribution, the diffraction line is weakened when $\rho_d^{\Sigma} < 10^{10}\ \text{cm}^{-2}$, at a larger value of $10^{10} \le \rho_d^{\Sigma} \le 10^{11}\ \text{cm}^{-2}$ is broadened, accepting a Gaussian form, and if $\rho_d^{\Sigma} > 10^{11}\ \text{cm}^{-2}$, that it is smeared, leaving fluctuation of diffuse scattering.

Structural inhomogeneity of a polycrystalline system affects the distribution of the scattering intensity near the diffraction maximum, manifesting itself in the slow weakening of the harmonics that represent its shape.

2. The role of fluctuations in the density of dislocations over randomly oriented crystals. Fluctuations in the dislocation density caused by dispersion of the crystals orientations can in general be comparable with purely random fluctuations in a polycrystal. Under this assumption, the number of defects in crystals of each of the structural components of a polycrystalline system is modeled by independent random variables with the same Poisson distribution. To reveal the actual effect of a dispersion of the crystals orientations, it is required to consider the joint distribution of a random number of defects simultaneously in all crystals of each of the subsystems.

In the approximate representation, the multidimensional probability distribution of a random number of defects **Π** in an ensemble of deformed crystals will be a product of independent Poisson distributions with random parameters.

If n_j is a parameter of the Poisson function for the j-th crystal, then the mean and variance of a random number of defects therein can be written as

$$\left\langle \Pi_j \right\rangle = n_j, \quad \left\langle \left| \Pi_j - \left\langle \Pi_j \right\rangle \right|^2 \right\rangle = n_j.$$

In the model of homogeneous structure all n_j (j = 1, 2, …) are the same. In reality, due to the dispersion of crystals orientations, the parameters n_j themselves are subject to random fluctuations:

$$n_j = \bar{n} + v_j,$$

where $\bar{n}$ is the average value of n_j over the ensemble of crystals, v_j are random deviations from $\bar{n}$, supposedly independent of each other and having the same probability distribution with mathematical expectation $\langle v_j \rangle = 0$ and variance q^2.

The expected mean number of defects fallen at one crystal and the expected mean square fluctuation of the number of defects over crystals are obtained in form [105]:

$$\overline{\langle \Pi \rangle} = \bar{n}, \quad \left\langle \left| \Pi_j - \overline{\langle \Pi \rangle} \right|^2 \right\rangle = \bar{n} + q^2.$$

From this it is clear that q^2 is the increment in the mean square of the fluctuations above the mean Poisson variance equal to $\bar{n}$.

Fourier representation of the diffraction intensity distribution is related to the function G of shape of the reciprocal lattice site. The joint probability distribution of the involved therein quantities $e^{-W\Pi_j}$ ($j = 1, 2, \dots, r$), where r is the volume of the scattering aggregate of crystals, is completely determined by the distribution law of the vector **Π**. Under assumption that all of its random components $\{\Pi_j\}$ have independent distributions, the functions $e^{-W\Pi_j}$ are mutually independent.

Mathematical expectation of the sum of r random functions $e^{-W\Pi_j}$ when r is an independent random variable with a Poisson distribution law is calculated using the generating function of a complex Poisson distribution [25].

Taking as the probability distribution of the vector **Π** a product of the Poisson distributions of the components $\{\Pi_j\}$, when $W \ll 1$ we find

$$\left\langle \sum_{j=1}^{r} e^{-W\Pi_j} \right\rangle \cong \langle r \rangle \int_{-\infty}^{\infty} e^{-W\left(\overline{\langle \Pi \rangle} + v\right)} \varphi(v)\, dv.$$

Here, $\langle r \rangle$ is the expected (average) number of scattering crystals; $\varphi(v)$ is the probability density of random fluctuations of the parameters in the function of the multidimensional Poisson distribution of the vector **Π**.

In all crystals of the ensemble under consideration, the number of defects is large. Great deviations from the general mean $\overline{\langle \Pi \rangle}$ are unlikely, consequently $\varphi(v)$ decreases rapidly with $|v|$. For a sufficiently large ensemble of crystals as an approximation to $\varphi(v)$ one can take the density of the normal (Gaussian) distribution:

$$\varphi(v) \cong \frac{1}{\sqrt{2\pi}q} e^{-\frac{v^2}{2q^2}}.$$

As a result, the equation of normalized harmonics of the diffraction line, with allowance for microinhomogeneity of a polycrystal, will be following:

$$A_k \cong e^{-W\langle\Pi\rangle + W^2 q^2/2}. \tag{8.3}$$

Parameters of Eq. (8.3) are related to the ensemble average of the dislocation density $\overline{\rho_d}$ and the additional mean-square fluctuations $\overline{|\Delta\rho_d|^2}$ caused by the inhomogeneity Poisson distribution of defects in ensemble:

$$q/\langle\Pi\rangle = \sqrt{\overline{|\Delta\rho_d|^2}} \Big/ \overline{\rho_d}.$$

Relative standard deviations, that is, the coefficients of variation denoted below by χ, are a measure of the dispersion of dislocation densities by randomly oriented crystals.

Figure 8.2 shows the effect of the dislocation density distribution in a polycrystal on harmonics of the diffraction line, taken in earlier as example.

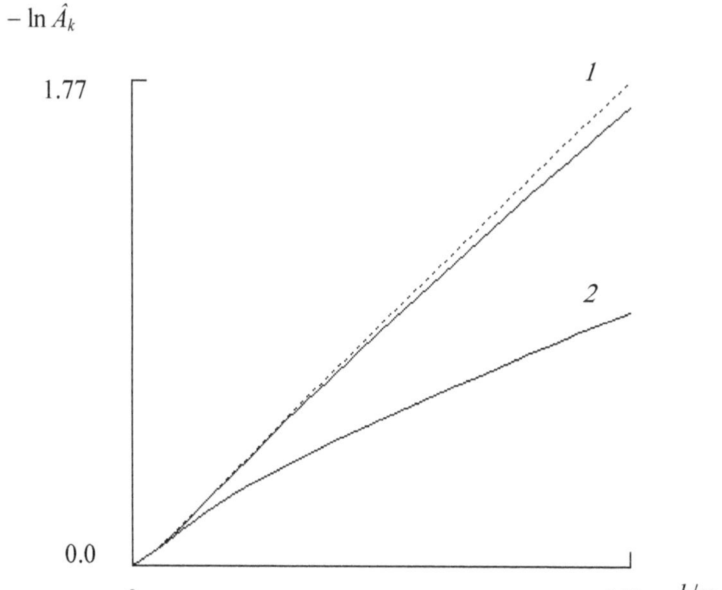

Fig. 8.2. Curves of logarithms of the calculated harmonics of the diffraction line depending on structural state of a polycrystal: (*1*) microinhomogeneity; (*2*) macroscopic inhomogeneity. Dashed line for uniform distribution of defects

For comparison there is taken a faintly inhomogeneous mixed aggregate of crystals $\left(\rho_d^{(1)} = 10^9\,\text{cm}^{-2},\ \rho_d^{(2)} = 10^{10}\,\text{cm}^{-2};\ \gamma_2 = 0.1\right)$ and a microinhomogeneous with the same total number of the defects $\left(\overline{\rho_d} = 2.3\cdot 10^9\,\text{cm}^{-2},\ \chi = 0.25\right)$.

With the same average dislocation density fallen at one crystal, the difference between logarithms of harmonics in the model of a mixed structure by Eq. (8.1) and in the model of a microinhomogeneous structure by Eq. (8.3) grows approximately in proportion to *k*.

Reduction of the decrease rate of harmonics with an increase in their order in the presence of a macroscopic fluctuation of the dislocation density in a polycrystal proves to be essential.

§ 8.2. Measuring of the Dislocation Density in an Inhomogeneously Deformed Polycrystal

The average and dispersion of dislocation density in each structural component of a polycrystalline system are under determination. At variety of information about the object a multivariant estimation of the parameters performs.

1. Model of diffraction observations for a mixed dislocation structure of a polycrystal. With a nonuniform dislocations distribution both within crystals and over crystals and structural components there is a theoretical model of diffraction observations on Eq. (8.1) – (8.3).

The model equation as applicable for measuring of parameters takes the following form:

$$\left.\begin{aligned} &A_k \cong f\left(x_k, \boldsymbol{\Theta}\right) + u_k, \\ &f\left(x_k, \boldsymbol{\Theta}\right) = \sum_{\nu=1}^{s} \exp\left[\varphi_\nu + \psi\left(x_k, \boldsymbol{\theta}_\nu\right)\right], \\ &\psi\left(x_k, \boldsymbol{\theta}_\nu\right) = \exp\left[-x_k\, e^{(\theta_0+\theta_1)_\nu} + \frac{1}{2} x_k^2\, e^{2(\theta_0+\theta_2)_\nu}\right], \\ &x_k = C_1\, Q_{HKL}\left(k/æ\right). \end{aligned}\right\} \qquad (8.4)$$

Here, x_k is the independent variable for reflection $\{HKL\}$; u_k is an auxiliary variable that compensates for systematic errors as in the model and observations data.

The parameters vector $\boldsymbol{\Theta} = \left(\boldsymbol{\Theta}_1, \ldots, \boldsymbol{\Theta}_s\right)^{t}$ has components $\boldsymbol{\Theta}_\nu = \begin{bmatrix} \varphi_\nu \\ \boldsymbol{\theta}_\nu \end{bmatrix}$ that contain the parameters $\boldsymbol{\theta}_\nu$ of the dislocation structure of the ν-th subsystem of crystals. The weight fraction of the ν-th component in the scattering crystals is represented by the parameter $\varphi_\nu = \ln\gamma_\nu$.

In the model of a one-component polycrystalline system will be the vector $\mathbf{\Theta} \equiv \mathbf{\theta}$.

The measuring parameters of the dislocations density distribution are contained in the components (θ_1, θ_2) of each vector $\mathbf{\theta}_\nu$ $(\nu = 1, \ldots, s)$:

$$\theta_1 = \alpha + \ln\overline{\rho_d},\ \ \theta_2 = \alpha + \frac{1}{2}\ln\overline{|\Delta\rho_d|^2},\ \ \alpha = \ln\left[a^2/2\pi\right].$$

Component θ_0 in all vectors $\mathbf{\theta}_\nu$ is the same. An "over" parameters of the dislocation structure are involved in it from W by Eq. (8.2), averaged over the subsystems:

$$\theta_0 = \ln\left\{\left(\bar{\xi}/a\right)\left[1+\left(\sigma_\xi/\bar{\xi}\right)^2\right]\left[1+(2\pi/p)\left(\bar{\tau}/b\right)^2\right]\right\},$$

which are only limited in the range of allowable values when optimizing the model Eq. (8.4).

Given form of function $f(x_k, \mathbf{\Theta})$ is adapted to the use of a priori information on the fractions of the structural components in an aggregate of the scattering crystals. Herein the parameters φ_ν are subject to equality type constraints $\varphi_\nu - \hat{\varphi}_\nu = 0$. If such information is not available, $f(x_k, \mathbf{\Theta})$ will contain the searched fractions of $\gamma_\nu = e^{\varphi_\nu}$ with imposed inequalities $0 < \gamma_\nu < 1$. In common there is the connection equation $\sum_{\nu=1}^{s}\gamma_\nu = 1$.

2. Estimation of parameters under conditions of systematic deviations of the observations data. Let us assume that the data of diffraction observations constitute a sample $\{\mathbf{A}_r\}$ $(r = 1, \ldots, n)$ of independent and identically distributed random variables of dimension K with a covariance matrix

$$\mathbf{V_A} = \begin{bmatrix} \sigma_1^2 & \cdots & 0 \\ \vdots & \ddots & \vdots \\ 0 & \cdots & \sigma_K^2 \end{bmatrix},$$

where $K \equiv k_{max}$ is the highest order of well-defined harmonics in the infinite-dimensional vector $\mathbf{A} = \{A_k\}$. It can be assumed that disturbances of the normal distribution law on sample are small.

Existing systematic deviations of the observations vector $\mathbf{A}$ from the model equations are most successfully approximated by trigonometric polynomials. Due to the orthogonality of the basis functions, there is no correlation between the error of approximation of the auxiliary variable $\{u_k\}$ and the error of estimation of $\mathbf{\Theta}$ by the maximum likelihood method [105].

Let $\overline{\varepsilon_k}$ be the sample mean of the deviations from the model

$$\varepsilon_{k,r} = A_{k,r} - f(x_k, \mathbf{\Theta}) \ (r = 1, \ldots, n).$$

Since the expected values of $\left|u_k - \overline{\varepsilon_k}\right|$ are zero, an acceptable approximation of $\{\tilde{u}_k\}$ is obtained when $\sum_{k=1}^{K}\left|u_k - \overline{\varepsilon_k}\right|^2 \Rightarrow \min$,

$$\tilde{u}_k = \frac{1}{K}\sum_{k'=1}^{K}\overline{\varepsilon_{k'}}\left[1 + 2\sum_{t=1}^{q}\cos t\omega_k \cos t\omega_{k'}\right], \quad \omega_k = \frac{\pi k}{K},$$

q is order of trigonometric polynomial.

A valid estimate of parameters vector $\mathbf{\Theta}$, which maximizes the likelihood of sample $\{\mathbf{A}_r\}$ with an inhomogeneous Gaussian distribution, is the stationary point $\mathbf{\Theta}^*$ of the objective function

$$L(\mathbf{\Theta}) = \frac{1}{2}\sum_{k=1}^{K}\sum_{r=1}^{n}\left|\varepsilon_{k,r} - \tilde{u}_k\right|^2 / \sigma_k^2. \quad (8.5)$$

The smallest order q of the polynomial for approximating $\{u_k\}$ is chosen, in which agreement of the model with the observations data is not rejected by statistical checks.

Criterion to checking the adequacy of the model by residual deviations of $\varepsilon^*_{k,r} = \varepsilon_{k,r}(\mathbf{\Theta}^*)$:

$$\eta = \zeta\sum_{k=1}^{K}\left\{\frac{1}{n}\sum_{r=1}^{n}\left|\varepsilon^*_{k,r} - \tilde{u}^*_k\right|^2 / \sigma_k^2\right\}, \quad \zeta = \left[1 - \frac{3s + q + 1}{nK}\right]^{-1}$$

by a normal distribution of measurement errors has an χ^2-distribution with K degrees of freedom. There is taken into account the correction ζ for bias of the variance estimate over residuals depending on the number of adjustable parameters in K equations of the model (8.4).

In practice, a good approximation to the systematic distortions $\{\hat{u}_k\}$ turned out to be a polynomial of order $q = 1$ for a one-component system ($s = 1$) and $q = 2$ for a two-component system ($s = 2$).

Parameters $\boldsymbol{\theta}_\nu$ ($\nu = 1, \dots, s$) must reside to the region of allowable values of $(\underline{\boldsymbol{\theta}}, \overline{\boldsymbol{\theta}})$, according to available information on the dislocation structure of plastic deformation (§ 7.2).

To restrict the search for a stationary point $\boldsymbol{\Theta}^*$ in the allowable region of the parameter space of the object, the penalty function is introduced:

$$Z(\boldsymbol{\Theta}) = -\lambda \sum_{l=0}^{2s} \left[\ln w_l + \ln(1 - w_l) \right], \tag{8.6}$$

where $\mathbf{w}$ is the vector of normalized parameters changing within the interval (0, 1):

$$w_l = \begin{cases} (\theta_0 - \underline{\theta_0}) / (\overline{\theta_0} - \underline{\theta_0}) & (l = 0), \\ (\theta_1 - \underline{\theta_1})_\nu / (\overline{\theta_1} - \underline{\theta_1})_\nu & (l = \nu = 1, \dots, s), \\ e^{2(\theta_2 - \theta_1)_\nu} & (l = s + \nu = (s+1), \dots, 2s). \end{cases}$$

Positive coefficient λ rapidly decreases during optimization. Its initial value is taken to be $\lambda^0 \sim 10^{-3} L(\boldsymbol{\Theta}^0)$, where $\boldsymbol{\Theta}^0$ is the allowable starting point [2].

From the quantitative analysis of the texture, it is possible to obtain approximate limitations for the component $\boldsymbol{\varphi}$ of parameters vector $\boldsymbol{\Theta}$: $g_\nu = \varphi_\nu - \tilde{\varphi}_\nu$, where $\tilde{\varphi}_\nu = \ln \tilde{\gamma}_\nu$ is calculated from the measured values of γ_ν ($\nu = 1, \dots, s - 1$). About the covariance matrix of the restrictions vector $\{g_\nu\}$ it is known only that it is proportional to the unit matrix: $\sigma^2_{\gamma_\nu} \sim \gamma^2_\nu$, so means $\sigma^2_{\varphi_\nu} \sim 1$, and hence $\sigma^2_{g_\nu} \sim 1$.

When assuming that the distribution of errors g_ν is not very different from the normal one, to the logarithm of the likelihood function in the form of Eq. (8.5) it is added the following quantity [2]:

$$H(\boldsymbol{\Theta}) = \frac{s-1}{2} \ln \sum_{\nu=1}^{s-1} |g_\nu|^2. \tag{8.7}$$

In the absence of measurements data of $\tilde{\gamma}_\nu$ the restrictions on the weight fractions of γ_ν will be controlled by the additional term of the penalty function by Eq. (8.6):

$$Z'(\boldsymbol{\Theta}) = -\lambda \sum_{\nu=1}^{s-1} \left[\ln \gamma_\nu + \ln(1 - \gamma_\nu) \right]. \tag{8.8}$$

Constructed optimization functional with constant members (8.5), (8.6) and additional (8.7) or (8.8) for a mixed structure, is effectively minimized by the Marquardt-Gauss-Newton method. In the course of computational experiments with a random selection of starting points from an allowable region, a sample of the measured values of the parameters vector $\boldsymbol{\Theta}$ is extracted sufficient to construct self-correcting confidence intervals for the initial structural parameters (§ 7.1).

3. Tests of the method for measuring the inhomogeneous dislocation density on model polycrystals. In the model of two-component structure of a polycrystal, on which the tests are carried out, there is assigned a mixed dislocation density distribution in different fractions (γ_1, γ_2) with parameters

$$\left(\bar{\rho}_d^{(1)} = 10^9\,\text{cm}^{-2},\ \chi_1 = 0.15\right);\ \left(\bar{\rho}_d^{(2)} = 10^{11}\,\text{cm}^{-2},\ \chi_2 = 0.05\right).$$

Measurements of the harmonics of the diffraction line {112} bcc (Fe) were modeled by normally distributed random variables with calculated mathematical expectation $\hat{A}_k$ (æ = 0.1) and one the same standard deviations $\sigma = 0.01$. The dimension of the observations vector K is limited by the requirement for harmonics $\hat{A}_k$ to be significant with a confidence probability $P = 0.99$.

Simulation measurements data and deviation of the approximate function $f(x_k, \boldsymbol{\Theta})$ from the exact mathematical expression used in the data generator are shown in Fig. 8.3.

Estimates obtained by different methods for the parameters of the dislocation density distribution in mixed aggregates of crystals are presented in Table. 8.1.

Approximate confidence intervals are constructed from independent random samples with the size of 60 measured values of the vector $\boldsymbol{\Theta}^*$. Average intervals are given over repeated 60 samples.

Table 8.1.

Parameters of dislocation density distribution in the model of two-component polycrystalline system

The weight fractions taken	*Approximate 90% confidence intervals of estimates**			
	Mean, $\bar{\rho}_d^{(\nu)}$ [cm^{-2}]		*Variation coefficient,* χ_ν	
	$(\hat{\rho}_d = 10^9)_{\nu=1}$	$(\hat{\rho}_d = 10^{11})_{\nu=2}$	$(\hat{\chi} = 0.15)_{\nu=1}$	$(\hat{\chi} = 0.05)_{\nu=2}$
$\hat{\gamma}_2 = 0.1$	$[1.0; 1.2]\cdot10^9$	$[0.6; 0.9]\cdot10^{11}$	[0.14; 0.18]	[0.010; 0.013]
	$[0.9; 1.2]\cdot10^9$	$[0.6; 0.8]\cdot10^{11}$	—	—
$\hat{\gamma}_2 = 0.5$	$[0.9; 1.2]\cdot10^9$	$[0.8; 1.1]\cdot10^{11}$	[0.16; 0.20]	[0.011; 0.014]
	$[0.8; 1.0]\cdot10^9$	$[0.6; 0.8]\cdot10^{11}$	—	—

* The first row under the restriction of estimate of the weight fraction γ_2 in the interval (0, 1); the second row when minimizing of discrepancy $|\gamma_1 - \hat{\gamma}_2|$ in approximation of homogeneous components.

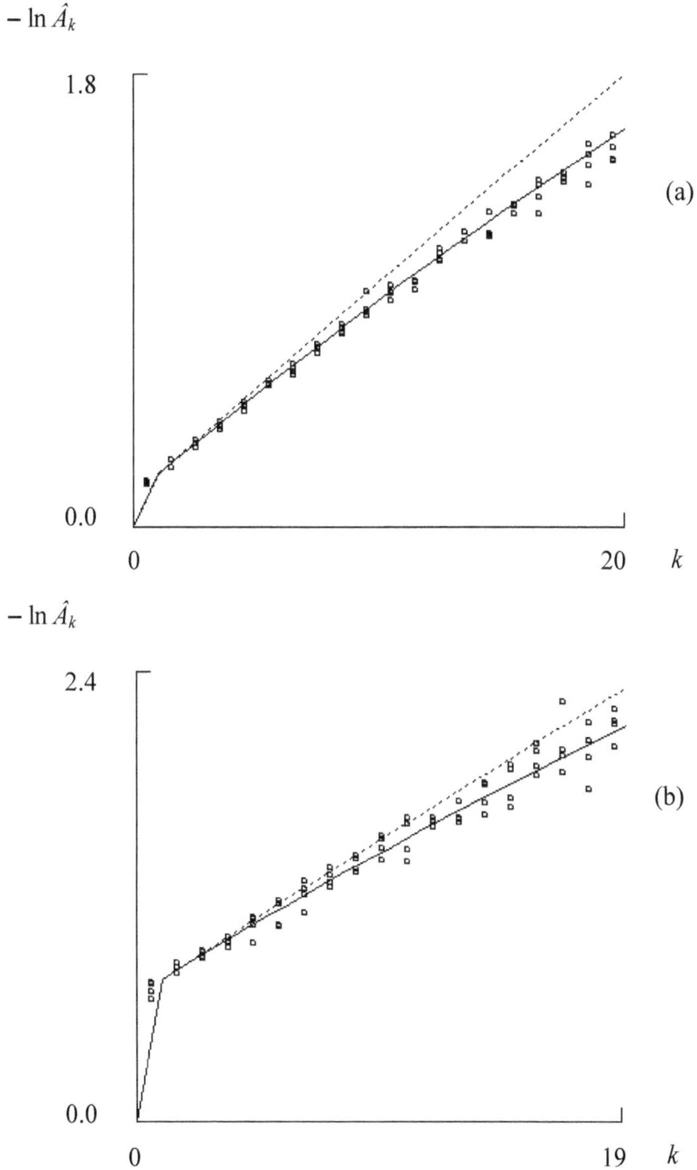

Fig. 8.3. Data of simulation measurements for an inhomogeneous aggregate of crystals: (a) $\gamma_2 = 0.1$; (b) $\gamma_2 = 0.5$. Solid curves present the exact diffraction equation, and dashed curves the approximation by model

Estimates of the means $\left(\overline{\rho}_d^{(1)}, \overline{\rho}_d^{(2)}\right)$ of mixed distribution of dislocation density are lightly biased to weighted average densities $\overline{\rho}_d^{\Sigma}$. Neglecting the fluctuations (χ_1, χ_2) lowers the estimates of average densities, and for large $\overline{\rho}_d^{\Sigma}$ the difference in estimates is more significant. Confidence intervals for the fluctuations (χ_1, χ_2), despite their arising bias, are close to the given values.

In principle, the method correctly distinguishes the dislocation densities even in weakly inhomogeneous aggregate of crystals.

§ 8.3. Parameters of Dislocation Density Distribution in Examples of Structures of Low-Carbon Steel

A macroscopic inhomogeneity of dislocation density on the components of crystallographic texture of low-carbon steel is reliably revealed. Fluctuations of dislocation density over randomly oriented crystals naturally weaken with an increase of the orientational order.

1. Density of dislocations into deformation texture. Probability distribution of the orientations of crystals in specimen of cold-worked low-carbon steel is determined from measured spherical harmonics (§ 2.3). Sampling distribution of orientations of a random aggregate of crystals is shown in Fig. 4.1. Texture parameters that are used in estimating the density of dislocations are given in Table 8.2.

Table 8.2.

Parameters of ordering of crystallographic planes in thin-sheet low-carbon steel

ν	$\{pqr\}_\nu$	μ_ν	K_ν
1	{112}	0.753 ± 0.004	20.3 ± 0.6
2	{111}	0.247 ± 0.004	42.3 ± 1.3

Table 8.3 shows the probabilities Γ_ν of the falling the crystals into the reflecting position when parameters of the orientation ordering are K_ν, as well as fractions of texture components in the diffraction doublet of γ_ν when their weight in a mixed distribution of crystals orientations is μ_ν ($\nu = 1, 2$). An angle of misorientations of the reflecting planes accepted for calculation is $\delta\vartheta = 0.1$.

On the set of reflections $\{HKL\}$ the texture components are best represented in the {222} reflection, but as the Γ value shows, the largest number of crystals can share in the reflection {112}.

Table 8.3.

Probability of reflection on orderly oriented crystals in specimen of thin-sheet low-carbon steel

Calculated quantity	*Reflection indices {HKL}*			
	{222}	{112}	{200}	{110}
$\Gamma_{\{112\}}$	0.0318	0.1086	0.0024	0.0134
$\Gamma_{\{111\}}$	0.2115	0.0565	0.0000	0.0003
$\Gamma^{(\mu_1,\mu_2)}$	0.0767	0.0956	0.0018	0.0101
$\gamma_{\{112\}}$	0.3108	0.8523	1.0000	0.9933
$\gamma_{\{111\}}$	0.6892	0.1477	0.0000	0.0067

Under conditions of existing texture, the predicted variance of harmonics of the diffraction line

$$\sigma_{A_k}^2 \sim \sum_{\nu=1}^{s} \mu_\nu \left(1-\gamma_\nu/\mu_\nu\right)^2$$

for {222} is an order of magnitude larger in comparison with {112}. With actual structural state of the specimen, the diffraction line {112} by reason of informative is more preferable for determining the density of dislocations.

X-ray measuring the test specimens from low-carbon steel has been performed by D.A. Kozlov.

The diffraction line {112} was measured in Co–K_α radiation in the interval $\Delta 2\theta_{\{112\}} = 10°$ with the step of 0.1° that satisfies the requirement of uncorrelated counts of pulses. X-raying was carried out in two passes with a counting time of 5 s per point at 72% strain of the specimen and 20 s at 18% strain, and then repeated with a new installation of the same as specimen. Reference line was measured on quite well annealed specimen. X-ray measuring conditions to obtain the most accurate information from the diffraction line are considered in § 9.1.

Optimal estimates of harmonics of the physical profile of measured diffraction lines are shown in Fig. 8.4. Method for determining the physical profile is described in § 9.2.

The presence of more than two structural components in a deformed polycrystal is possible, but it is practically impossible to reveal them when a limited amount of reliable information in the experimental data. A model of one-component or two-component structure of a polycrystal is chosen by the method of successive regression analysis of the diffraction observations $Y_k = -\ln A_k$.

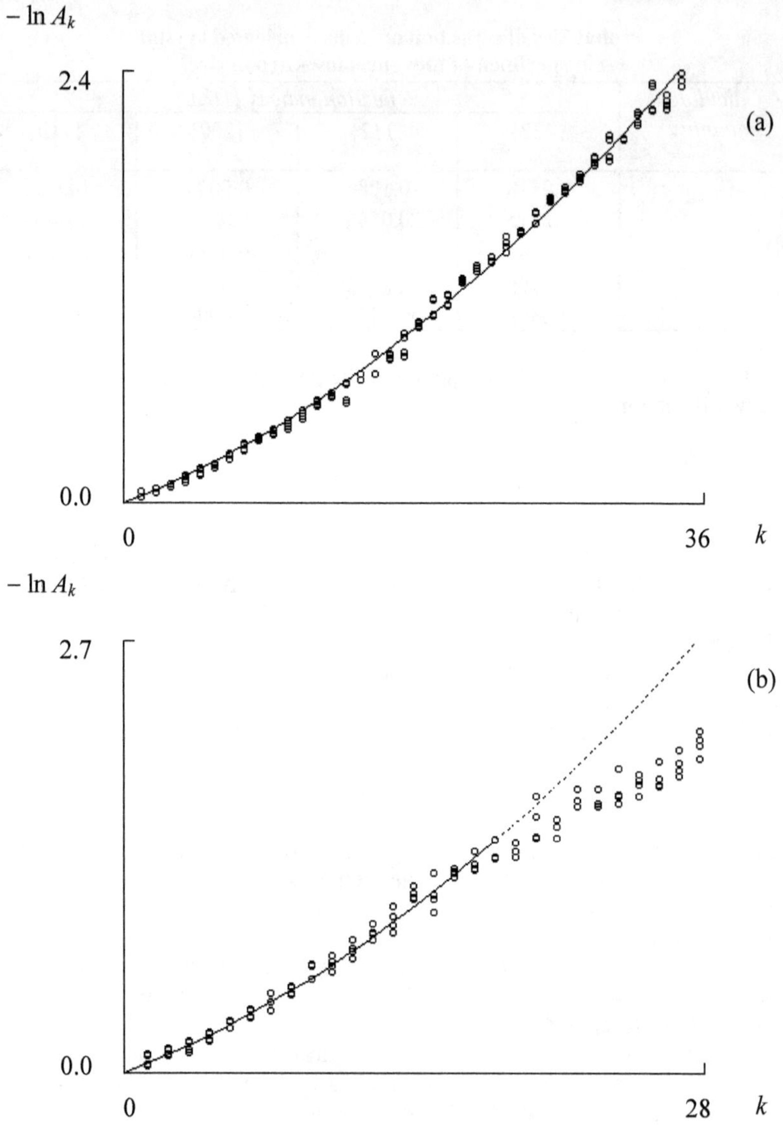

Fig. 8.4. Measured harmonics of the physical profile of the diffraction line of low-carbon steel samples with different extent of deformation: (a) 18%; (b) 72%

Table 8.4.

Estimates of the dislocations system parameters by a homogeneous polycrystal model for thin-sheet low-carbon steel

Approximate 90% confidence intervals of estimates			
Dislocations density $\overline{\rho}_d$	*Average radius of loops* $\overline{\xi}/a$	*Variation coefficient* $\sigma_{\xi}/\overline{\xi}$	*Correlation radius* $\overline{\tau}/b$
$[2.1; 2.6]\cdot 10^9$ cm^{-2} $[2.5; 3.2]\cdot 10^{9*}$ cm^{-2}	[460; 555] —	[0.20; 0.24] —	[1.5; 1.8] —

* With the restriction of other parameters in allowable region.

Table 8.5.

Average dislocation density in the components of the crystallographic texture of the low-carbon steel specimen

Approximate 90% confidence intervals	
{112} ⟨110⟩	{111} ⟨110⟩
$[2.0; 2.4]\cdot 10^9$ $(1.6-2.5)\cdot 10^9$ cm^{-2} $[1.9; 2.4]\cdot 10^{9*}$ cm^{-2}	$[2.2; 2.7]\cdot 10^{11}$ $(2.5-4.0)\cdot 10^{11}$ cm^{-2} $[2.2; 2.8]\cdot 10^{11*}$ cm^{-2}

* Without using information on the distribution of the orientations of crystals. (In parentheses these are most plausible estimates by the histogram maxima.)

Macroscopic inhomogeneity of a polycrystal is detected due to the fact that in the regression equations belonging to different models of the structure, the difference of output quantities follows $\sim k$ (Fig. 8.2). The basis for regression is constructed taking into account the available systematic deviations of the observations data caused by the asymmetry of the measured diffraction line[1] (§ 7.1).

Table 8.4 and 8.5 present the estimates of parameters of considered models of the object. On the results of computational experiments there is determined the average of confidence intervals for 900 random samples with the size of 60 extracted from total sample of $9\cdot 10^3$.

Wald's sequential criterion of likelihood ratio [2] shows the superiority of the model of a mixed dislocation structure of the specimen in comparison with the model of a homogeneous structure under an increase the highest order of the measured harmonics to $k \geq 23$. Probability to accept mistakenly both one and another model, when in fact the alternative is true, is by $P < 0.01$.

In the model of a homogeneous polycrystal, a "smoothed" structure is revealed with a dislocation density close to main in the specimen.

[1] Flattened shape of crystals in the plane-strained specimen creates an asymmetry along with the splitting of the doublet $K_{\alpha_1} - K_{\alpha_2}$ (§ 6.2).

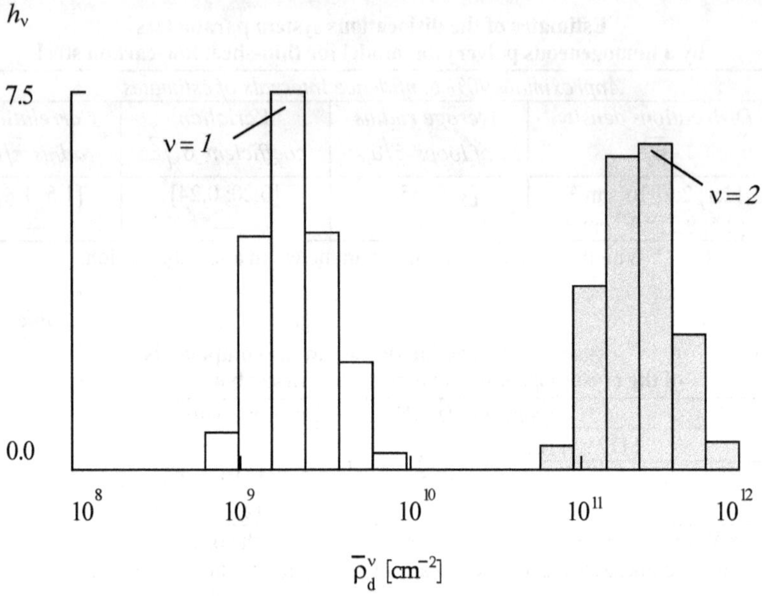

Fig. 8.5. Sampling distribution of the dislocation density measurements in the texture components of cold-worked low-carbon steel: (*1*) {112}⟨110⟩; (*2*) {111}⟨110⟩

As experience proves, estimates of the average density of dislocations in the structural components of a polycrystalline system are stable to a priori information on the weight fractions of the components in a scattering aggregate of crystals.

Sampling distribution of the measured values of a mixed dislocation density is represented by histograms combined in Fig. 8.5. The most plausible estimates of the densities from the histograms maxima are given in Table 8.5.

Reproducible components of the diffraction intensity distribution inherent in the crystallographic texture of the specimen are completely different by shapes, as it shown in Fig. 8.6.

On the basis of the results obtained, it can be concluded that in the texture component of cold-worked low-carbon steel with a high degree of ordering of {111}⟨110⟩ orientations, the dislocation density is two orders of magnitude higher than in the predominate component with the most probable orientation {112}⟨110⟩.

In a subsystem with a dislocation density above 10^{11} cm^{-2}, the disordering of the crystal lattice is so great that only a bell-shaped increase in the intensity of diffuse scattering remains on the place of the diffraction line, in full accordance with the theory of M.A. Krivoglaz.

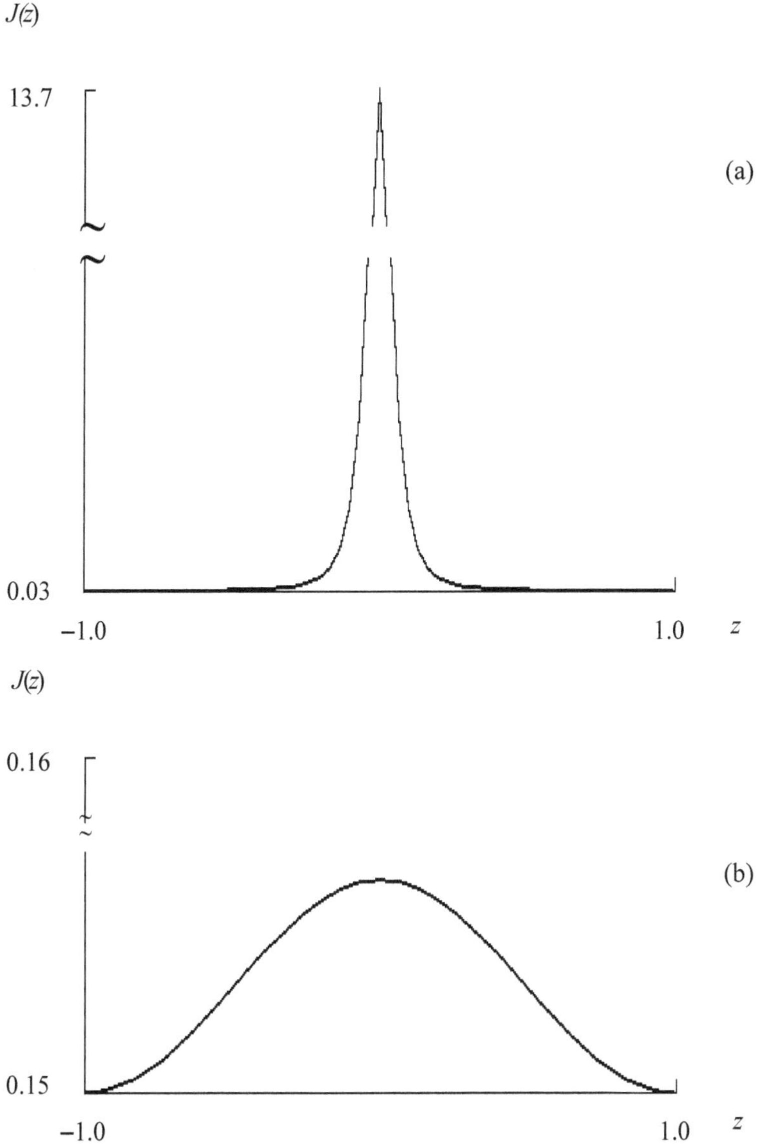

Fig. 8.6. Diffraction mapping the texture components of thin-sheet low-carbon steel: (a) {112}⟨110⟩; (b) {111}⟨110⟩

2. Microinhomogeneity of the dislocation density produced by the dispersion of orientations of crystals. Microinhomogeneous dislocation density was measured on a weakly deformed (18%) and strongly deformed (72%) specimens of low-carbon steel. The model of diffraction observations was chosen according to the structural states of the specimens, identified by successive regression analysis of the data: it is one-component for weakly deformed specimen and two-component for strongly deformed one.

Table 8.6 represents the estimates of the dislocation density distribution parameters in the structures of cold-worked low-carbon steel. Confidence intervals for mean densities of $\overline{\rho_d}$ and relative standard deviations of χ, so called variation coefficients, on average are turned out of 900 random samples of size 60, extracted from the total sample 9×10^3 of the measurement results. Sampling distribution of the measured values of the parameters is shown in Fig. 8.7.

Table 8.6.

Parameters of dislocation density distribution in structures of low-carbon steel with different degrees of ordering of crystal orientations

Measuring value	*Approximate 90% confidence intervals*		
	Short range order	*Long range order*	
		weak	*strong*
Average density, $\overline{\rho_d}$ [cm^{-2}]	$[1.5; 1.8]\cdot10^9$	$[2.5; 3.0]\cdot10^9$	$[3.0; 3.6]\cdot10^{11}$
Variation coefficient, χ	[0.19; 0.24]	[0.06; 0.07]	[0.005; 0.006]
Criterion for fluctuation significance, η	423	64.7	47.7

If the fluctuations in the dislocation density over the crystal orientations do not exceed its random fluctuations in crystals with one and the same orientation, then the following limitation will be satisfied with probability $P = 1 - \varepsilon$:

$$\eta = \overline{|\Delta\rho_d|^2} \Big/ \sigma^2_{\overline{\rho_d}} < u_r(\varepsilon),$$

where $\sigma^2_{\overline{\rho_d}}$ is the mean Poisson variance of dislocation density in a polycrystal; $u_r(\varepsilon)$ is percentage points of the gamma-distribution for sample of size r.

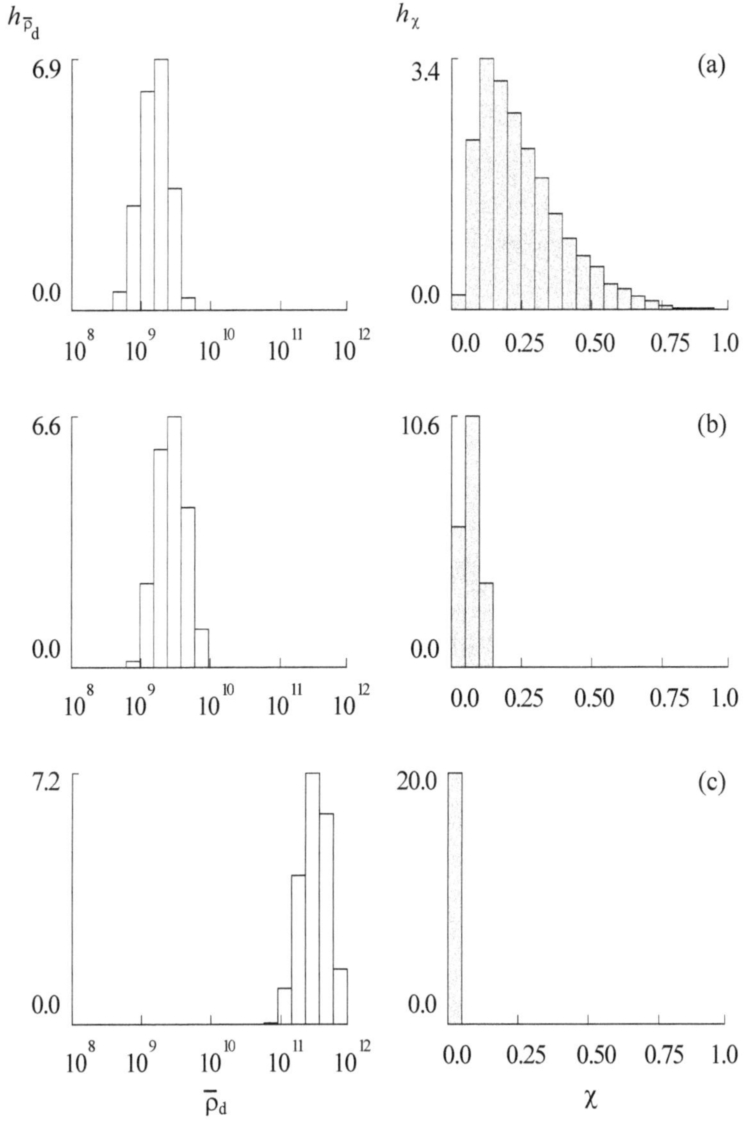

Fig. 8.7. Sampling distribution of the measurements of the mean and variations of dislocation densities ($\overline{\rho_d}$ [cm^{-2}], χ) in crystals with the ordered orientations: (a) short-range order; (b) low degree of long-range order of {112}⟨110⟩ type; (c) high degree of long-range order of the {111}⟨110⟩ type

Let us take the variance $\sigma^2_{\overline{\rho_d}} = \overline{\rho_d}\left(2\pi\overline{\xi}/V\right)$ for the greatest in the range of allowable values of the average radius of dislocations loops $\overline{\xi} \sim 1000\ a$ and the smallest observed crystal size $V^{1/3} \sim 10$ μm. The values of η for the largest of all possible $\sigma^2_{\overline{\rho_d}}$ are entered in Table 8.6.

From a comparison of the lowest evaluations of η with the critical value of the criterion at $\varepsilon = 0.0002$ [105]:

$$u_r(\varepsilon) = \begin{cases} 0.425, & r = 100; \\ 3.56, & r = 10; \\ 8.21, & r = 2 \end{cases}$$

it is clear that even for two scattering crystals the detected fluctuations are unlikely if the structure is homogeneous.

Microinhomogeneity of dislocation density proved to be significant even with strong ordering of the orientations of crystals. When neglecting by fluctuation over randomly oriented crystals, the estimates of the mean dislocation density in structural components biased to lower values.

The observed relation between a level of fluctuations in the dislocations density and state of order in orientations of crystals is consistent to physical conceptions. Consequently, in practice, the method yields realistic parameters of the complex dislocation structure of a polycrystalline system.

CHAPTER 9
OPTIMUM MEASUREMENTS OF DIFFRACTION ON POLYCRYSTALS FOR STRUCTURAL ANALYSIS

The recovery with the greatest possible accuracy of the true diffraction intensity distribution inherent in the object under study is the goal of optimal organization and processing of diffractometric data [69, 87–88].

§ 9.1. Fourier Representation of Measured Diffraction Line with Adaptation to a Background Level

At the adaptive method of search of the coefficients of Fourier series, representing the shape of the diffraction line, its background component refers to systematic errors in the data. The length of the Fourier series agrees with the variance of the measurements.

1. A self-adjusting model for estimating the harmonics of the diffraction line. Let us introduce the intensity distribution function $\varphi(x)$ for Bragg scattering that is merging with the diffuse scattering when it "thickens" toward the reciprocal lattice site [39]. A deterministic drift of diffuse scattering, which manifests itself in the growth of its intensity with distance from the zero point of the reciprocal space, is denoted by $\chi(x)$. Overall change in the scattering intensity within range of the reciprocal lattice period will be described by the function

$$f(x) = \varphi(x) + \chi(x).$$

Let $\mathbf{Y} = (Y_1, \ldots, Y_M)$ be a vector of diffraction measurements, where $Y_m = Y(x_m)$, and x_m are the normalized coordinates of M observation points in the interval $(-1 \le x \le 1)$. The problem is to determine the distribution function of the scattering intensity $\varphi(x)$ from the measurement data $\mathbf{Y}$.

To define a function $\varphi(x)$ means to estimate the coefficients of its Fourier representation:

$$\varphi(x, \mathbf{c}) \cong \sum_{k=-K}^{K} c_k e^{ik\pi x} \quad \left(K < (M-1)/2\right).$$

The vector of harmonics $\mathbf{c}$ carries physical information. Next section of this paragraph is devoted to the method of its optimal estimation.

Interfering drift is conveniently approximated by the orthonormal Legendre polynomials $P_l(x)$ [37]:

$$\chi(x, \mathbf{h}) \cong \sum_{l \ge 0} \sqrt{\frac{2l+1}{2}} h_l \, P_l(x).$$

From the regression model of the measured value:

$$Y_m \cong f(x_m, \mathbf{c}, \mathbf{h}) \quad (m = 1, \ldots, M) \tag{9.1}$$

it is required to exclude a purely fitting vector of coefficients **h**, to calculate of which the usual least-squares method is suitable.

Condition

$$\sum_{m=1}^{M} \left| Y_m - \varphi(x_m, \mathbf{c}) - \chi(x_m, \mathbf{h}) \right|^2 \Rightarrow \min$$

with an accuracy to $\sim M^{-1}$ there are satisfied the coefficients

$$h_l = \sqrt{\frac{2l+1}{2}} \left[\frac{1}{M} \sum_{m=1}^{M} P_l(x) Y_m - \sum_{k=-K}^{K} c_k i^l j_l(k\pi) \right].$$

Here, $j_l(r)$ are spherical Bessel functions of the first kind, which are expressed through trigonometric functions (sin r, cos r) [37]. In particular,

$$j_0(k\pi) = \begin{cases} 1, & k = 0, \\ 0, & k \neq 0; \end{cases} \quad j_l(0) = 0 \quad (l > 0);$$

$$j_l(k\pi) = \frac{(-1)^{k+1}}{k\pi} \begin{Bmatrix} 1, & l = 1, \\ 3/k\pi, & l = 2 \end{Bmatrix} \quad (k > 0).$$

Equation (9.1) when eliminating **h** becomes a self-adjusting model for estimating the vector of harmonics **c**:

$$\left.\begin{aligned} & Z_m \cong g(x_m, \mathbf{c}) \quad (m = 1, \ldots, M), \\ & Z_m = Y_m - \sum_{l \geq 0} \sqrt{\frac{2l+1}{2}} P_l(x_m) \left[\frac{1}{M} \sum_{n=1}^{M} P_l(x_n) Y_n \right], \\ & g(x_m, \mathbf{c}) = \sum_{k=-K}^{K} c_k \left[e^{ik\pi x_m} - \sum_{l \geq 0} \sqrt{\frac{2l+1}{2}} P_l(x_m) i^l j_l(k\pi) \right]. \end{aligned}\right\} \tag{9.2}$$

The number of degrees of freedom is decreased on the dimension of the vector **h**, which is one more than the highest degree of the polynomials $P_l(x)$. Between random errors **c** and **h** there is a correlation of order M^{-1}. Such is a growth coefficient for variance of estimates $\sigma_{c_k}^2$ as a "surcharge" for setting up.

Experience suggests that an acceptable representation of drift $\chi(x)$ will be given by $P_l(x)$ polynomials no higher than the first degree. Then in Eq. (9.2), the output variable Z_m and the regression function $g(x_m, \mathbf{c})$, where kind of the coefficients are $c_k = \frac{1}{2}\left(a_k - ib_k\right)$, take the form

$$\left.\begin{aligned} Z_m &= Y_m - \frac{1}{2M}\sum_{n=1}^{M} Y_n\left[1+3x_m x_n\right], \\ g(x_m,\mathbf{c}) &= \frac{a_0}{4} + \\ &+\sum_{k=1}^{K}\left\{a_k \cos k\pi x_m + b_k\left[\sin k\pi x_m - \frac{3}{2}\frac{\left(-1\right)^{k+1}}{k\pi}x_m\right]\right\}. \end{aligned}\right\} \quad (9.3)$$

Consequently, adaptive harmonic analysis requires very small changes of the basis functions of trigonometric regression.

2. Vector of harmonics of the diffraction line with the lowest generalized variance. In an infinite-dimensional space of Fourier representations, the estimate of the vector **c** is optimal if it bias is limited. Limitation of biases is possible in principle when the a priori information on the smoothness of the regression function is included and the error of the function is consistent with the data errors [95, 49].

Method of stable estimation of the Fourier coefficients of a diffraction line requiring restrictions on biases consists of the following actions:

- Iterative search of the minimum necessary length of the Fourier series to approximate the data: the higher order of harmonics K is increased until reducing of deviations from the regression function come to end.
- A choice from the set of vectors **c** for which the dispersion of data relative to the approximating function is comparable with their errors, of vector with the lowest norm $\|\mathbf{c}\|^2$.

For each K a vector **c** is searched out that corresponds to a minimum of the objective function of multidimensional linear regression

$$\left.\begin{aligned} Q(\mathbf{c}) &= \left[\mathbf{Rc}-\mathbf{Z}\right]^{\mathrm{t}}\mathbf{V}^{-1}\left[\mathbf{Rc}-\mathbf{Z}\right]+\omega\left[\mathbf{c}^{\mathrm{t}}\,\mathbf{L}\,\mathbf{c}\right], \\ \|\mathbf{c}\|^2 &< \left[\mathbf{c}^{\mathrm{t}}\,\mathbf{L}\,\mathbf{c}\right] < K^2\|\mathbf{c}\|^2. \end{aligned}\right\} \quad (9.4)$$

Here, **R** is the matrix made up of $(2K + 1)$ basis functions of the trigonometric regression by Eq. (9.3) that are calculated at the measurement points x_m $(m = 1, \dots, M)$; **L** is a diagonal matrix with elements of $L_{kk} = k^2\delta_{kk}$ $(k = 1, \dots, K)$ (t-superscript denotes transpose).

As an approximation to the covariance matrix of errors in the vector of experimental data **Z**, when $M \gg 1$, we can take

$$\mathbf{V} = \begin{bmatrix} \sigma_1^2 & 0 & \cdots & 0 \\ & \sigma_2^2 & & \vdots \\ & & \ddots & 0 \\ & & & \sigma_M^2 \end{bmatrix},$$

σ_m^2 is variance of primary measurements Y_m ($m = 1, \ldots, M$), which minimum value is determined by Poisson fluctuations of intensity.

A positive parameter ω regulates the extent of smoothing of the regression line according to the dispersion of data. To select the optimal ω^* by a minimum of $Q(\mathbf{c})$, it is used a geometric grid of type $\omega_j = \omega_0 d^j$, where $d > 0$ ($j = 0, 1, \ldots$) [95].

The measure of the error on the approximation of measurements with the total number of adjustable parameters $(2K + 3)$ it is

$$\eta = \frac{M}{M - (2K + 3)} \sum_{m=1}^{M} V_{mm}^{-1} \left[g(x_m, \mathbf{c}) - Z_m \right]^2.$$

Criterion of η initially decreases rapidly with increasing K, but in the following it is already stable growing. Hence, the residual deviations became comparable to the actual data errors [13].

When distribution of errors of **Z** is close to the normal, the value of η should not exceed strongly the critical value of the χ^2-criterion with M degrees of freedom. If, however, $\eta > M/(1 - P)$, the approximating equation is rejected as inadequate with the reliability not less of P.

When an acceptable agreement with the data is reached, the vector

$$\mathbf{c}^* = \left[\mathbf{R}^t \mathbf{V}^{-1} \mathbf{R} + \omega^* \mathbf{L} \right]^{-1} \left[\mathbf{R}^t \mathbf{V}^{-1} \mathbf{Z} \right]$$

is the estimate with the lowest generalized variance measured by determinant of the covariance matrix

$$\det \left[\mathbf{R}^t \mathbf{V}^{-1} \mathbf{R} + \omega^* \mathbf{L} \right]^{-1},$$

and with a minimum possible bias when the existing error of the regression function [2].

3. Testing the adaptive harmonic analysis of the diffraction line using simulation experiments. Simulation of diffraction measurements makes it possible to check the accuracy of the harmonics estimates by their deviations from true values.

Let us assume that the distribution of the scattering intensity is described by the Lorentz function appearing in the diffraction theory of deformed crystals [39]:

$$\varphi(x) = \left[1 + \lambda^2 x^2\right]^{-1}.$$

Drift of diffuse scattering (the cause of which in point-like perturbations of periodicity that exist even in an ideal crystal due to thermal vibrations of atoms), as results from the theory of [39], can obey the following relationship:

$$\chi(x) = \alpha\left[1 - e^{-(\beta x+1)^2}\right] \left(-1 \le x \le 1\right).$$

Let us agree to call the function $\chi(x)$ a background. When $\beta \ll 1$, as on reality, $\chi(x)$ approaches the equation of a straight line with an angular coefficient of $\sim(\alpha\beta)$.

Parameters (λ, α, β) are chosen so that the experimental characteristics: the peak width by proportion of the observation interval, the average relative level of a background and it's the slope angular rate were comparable to the real ones. In Fig. 9.1 it is depicted a profile of the measured function $f(x)$ with the parameters $\lambda = 6\pi$, $\alpha = 0.30$, $\beta = 0.03$.

Measurements data at a given number of points M of the interval ($-1 \le x \le 1$) are modeled by mutually independent Poisson random variables ($Y_1, \ldots, Y_M$), whose distribution is close to the normal one with the mathematical expectation $f(x_m)$ and the variance $\sigma_m^2 \sim f(x_m)$ ($m = 1, \ldots, M$). Regulating ratio on Poisson dispersion under simulation measuring is consistent with the scale of real intensities of diffraction.

In the accepted practice of primary data processing, the extreme points of the observations interval that fit into the overall regression line taken as count out level are considered to be background and discarded. Let us call usual way the "background cutting-off".

In Table 9.1 there are presented estimates of the beginning harmonics c_k for different methods of analyzing measurements in comparison with the exact values of $\hat{c}_k$ from the integral Fourier transform $\varphi(x)$ ($-\infty \le x \le \infty$). There is given the average of 99% confidence intervals, constructed from the results of $N = 60$ repeated experiments, with the number of measurement points $M = 121$. The measurements variance is $\sigma_m^2 / f(x_m) = 0.006$ ($m = 1, \ldots, M$).

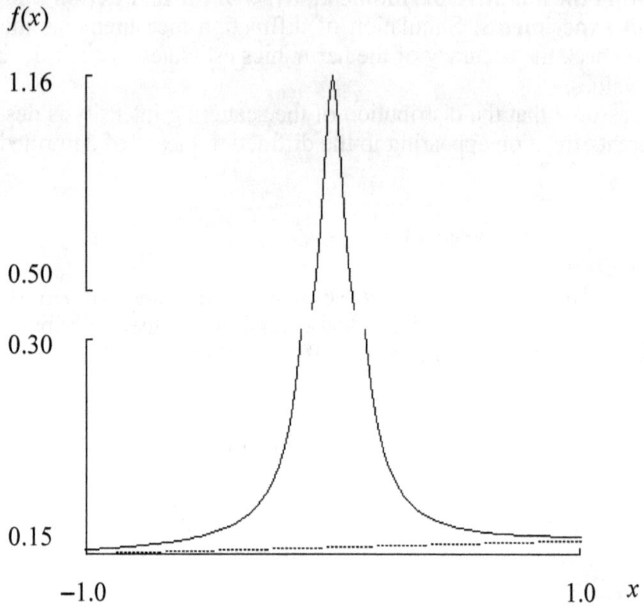

Fig. 9.1. Model function for simulation tests of adaptive harmonic analysis. Dashed line depicts the background function of $\chi(x)$

Table 9.1.

Estimates of the beginning harmonics of the model profile from simulation measurements data

k	*Exact harmonics* $\hat{c}_k$	*Method of analysis of the experiment*	
		adaptation to background	*cutting-off a background*
1	0.8465	[0.8551; 0.8918]	—
2	0.7165	[0.7190; 0.7500]	[0.7946; 0.8144]
3	0.6065	[0.6097; 0.6376]	—
4	0.5134	[0.5133; 0.5389]	[0.5591; 0.5746]
5	0.4346	[0.4345; 0.4583]	—

Figure 9.2 shows how the deviation of the averaged estimates of c_k from the exact theoretical harmonics $\hat{c}_k$ increases at $k \to 0$. With the same regularity there are deviated the coefficients of approximation $\varphi(x)$ by trigonometric polynomial on interval $(-1 \le x \le 1)$, these will be called approximate harmonics $\tilde{c}_k$.

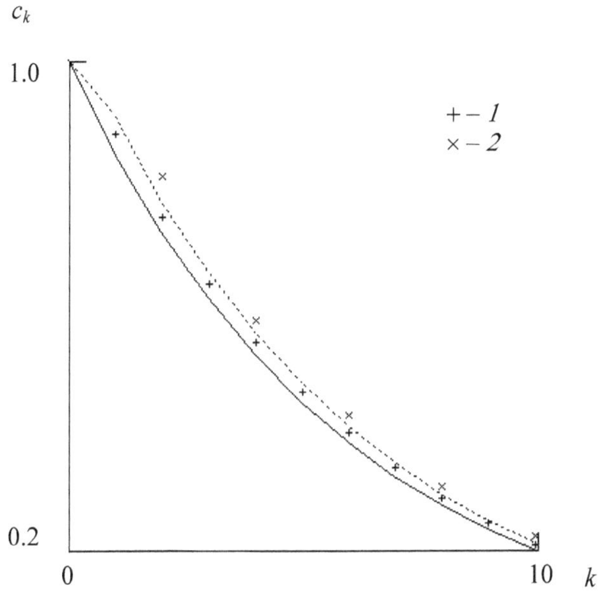

Fig. 9.2. The beginning harmonics of model diffraction profile: solid line is exact harmonics ($-\infty \le x \le \infty$); dashed line is approximate harmonics ($-1 \le x \le 1$). Points denote the average of estimates from a series of simulation experiments: (*1*) adaptation to background; (*2*) cutting-off a background

Confidence intervals of self-correcting estimates of c_k are biased from the exact harmonics of $\hat{c}_k$ to an approximate high $\tilde{c}_k$. The "background cutting-off" causes over-estimates of the beginning harmonics already with respect to $\tilde{c}_k$ in addition to the large loss of information contained in the diffraction measurements.

With an existing bias, the measure of the loss in accuracy of estimate c_k is the distance of upper confidence boundary from the true harmonic [105]:

$$E(P) = \left[c_k + \sigma_{c_k} t\big|_{1-P} \right] - \hat{c}_k,$$

where σ_{c_k} is the standard deviation, $t\big|_{1-P}$ is the percentage point of the t- distribution for the probability P. In Table 9.2 it is given the value of $E(P)$ for first-order harmonic most loosing in the accuracy of estimate. Probability that its error with different original data will not exceed the specified values, is $P = 0.99$.

Table 9.2.

Limit losses in accuracy of estimates of the first order harmonic

The measurements variance $\sigma_m^2/f(x_m)$	*The parameters of simulation experiments*			
	$M = 121$, $\lambda = 6\pi$		$M = 61$, $\lambda = 3\pi$	
	$\alpha = 0.00$, $\beta = 0.00$	$\alpha = 0.30$, $\beta = 0.03$	$\alpha = 0.00$, $\beta = 0.00$	$\alpha = 0.30$, $\beta = 0.03$
0.002	0.029	0.037	0.051	0.056
0.006	0.032	0.044	0.053	0.061

To reduce the deviation of c_k from the true harmonics $\hat{c}_k$ it is required the greater decrease in the measurements variance when the higher the background parameters $\chi(x)$. An increase of the highest degree of polynomials $P_l(x)$ to the model Eq. (9.2) does not improve the accuracy of estimates c_k, even with a simultaneous reducing the variance of the measurements by an order of magnitude.

Achievable accuracy of the beginning harmonics of the diffraction line is controlled by length of the interval of observations. Quality of the higher harmonics depends most on the method of constructing the regression equation approximating the measurement data.

In the traditional method of least squares, the length of the Fourier series is limited by the number of measuring points, and in the method of steady regression analysis, one is consistent with data errors. So that appearing error of the regression function varies.

To check the accuracy of the methods there have been calculated the root-mean-square deviations of the estimates of c_k from the true values of $\hat{c}_k$ over sampling of size $N = 125$ obtained as a result of simulation experiments on the model function $f(x)$ with parameters ($\lambda = 6\pi$, $\alpha = \beta = 0$). A criterion for the sufficiency of the length of a Fourier series for approximating the measurement data is $\eta < M$.

Figure 9.3 shows the growth curves of the true relative errors in estimates of high order harmonics, precisely

$$r_k = \frac{1}{\hat{c}_k}\sqrt{\overline{\left|c_k - \hat{c}_k\right|^2}},$$

when dispersion of measurements $\sigma_m^2/f(x_m) = 0.002$ ($m = 1, \ldots, 121$).

In Table 9.3 the greatest relative biases $\left(\overline{c}_k - \hat{c}_k\right)/\hat{c}_k$ of sample mean $\overline{c}_k$ for different accuracy of measurements are compared.

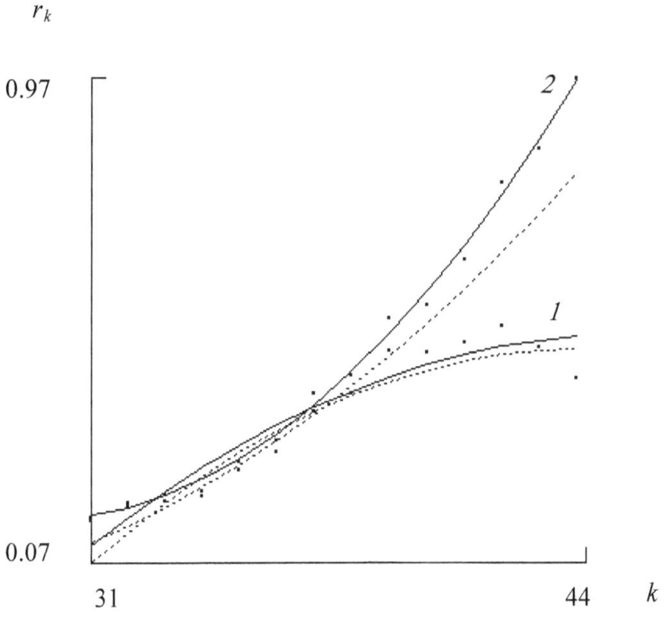

Fig. 9.3. Relative standard deviations of estimates from true harmonics:
(*1*) steady estimation; (*2*) method of least squares.
Dashed line for deviations from sample mean

Table 9.3.

Relative errors of the higher-order harmonics depending on variance of simulation measurements

Calculated ratio to the $\hat{c}_k$	$\sigma_m^2 = 0.002 f(x_m)$ $K = 44$		$\sigma_m^2 = 0.006 f(x_m)$ $K = 39$	
	1	2	*1*	2
Bias of the mean of estimates	−0.17	0.58	−0.33	0.41
Standard deviation of estimates	± 0.42	± 0.97	± 0.48	± 0.73

(*1*) method of steady estimation; (*2*) method of least squares.

A consequence of a finite number of measurement points M are positive biases of harmonics estimates [106]. With the increase in error of the regression function due truncate a series length of K, negative biases of high order harmonics grow rapidly. In spite of rising bias, the largest true relative error r_k of the stabilized estimate c_k ($k = K$) has remained at a lower level in the tests carried out.

Simulation experiments verify that in the adaptive method of approximation of measurements the beginning harmonics of the diffraction line are determined with the best accuracy. Efficiency of the method of steady estimation in comparison to the least squares method for higher-order harmonics increases significantly with increasing data accuracy.

4. Applying of the adaptive method of harmonic analysis to the data of a real experiment. Self-correcting on background, harmonics of the diffraction line $\{c_k\}$ are estimated by the normalized primary measurements of the scattering intensity

$$y_m = \frac{Y_m - Y_0^{\min}}{Y_0^{\max} - Y_0^{\min}} \quad (m = 1, \dots, M).$$

As count-out level of $Y_0^{\min}$, located at the point $x = 0$, the average for the extreme pairs of points of the observation interval is taken, the center of which is coincident with the maximum of the diffraction line of $Y_0^{\max}$.

A background component of the measured intensities is approximated by equation

$$\chi(x_m) \cong \frac{1}{2M} \sum_{n=1}^{M} \left[y_n - \sum_{k=-K}^{K} c_k e^{ik\pi x_n} \right] (1 + 3\, x_m x_n).$$

By the experimental conditions, the primary observations $\{Y_m\}$ are independent asymptotically normally distributed quantities with mathematical expectation $\hat{Y}_m = \hat{Y}(x_m)$ and variance $\sigma_m^2 = \hat{Y}_m / t$. Measuring time t is the same for all points of x_m. Diagonal covariance errors matrix of the normalized variables y_m has elements $V_{mm} = \sigma_m^2 / \left[Y_0^{\max} - Y_0^{\min} \right]^2$ $(m = 1, \dots, M)$.

For a practical example used measurement data of the diffraction line {112} of low-carbon steel specimen with deformation of 32%, which were obtained by D.A. Kozlov. X-raying procedure is detailed in § 8.3.

Poisson variance of the measured normalized intensities of y_m this is $V_{mm} = 0.004\,(y_m + 0.23)$. The given measuring points number is $M = 101$.

Sampling fluctuations of the Fourier coefficients are shown in Fig. 9.4. Profile of the diffraction line, adequate onto original data, which was constructed from the results of adaptive harmonic analysis, is presented in Fig. 9.5.

Method developed, in practice carries out the Fourier representation of the measurements corresponding to their accuracy. Harmonics of the diffraction line are determined directly from the primary data. Probability that these contain same amount of information as is obtained in the experiment rises.

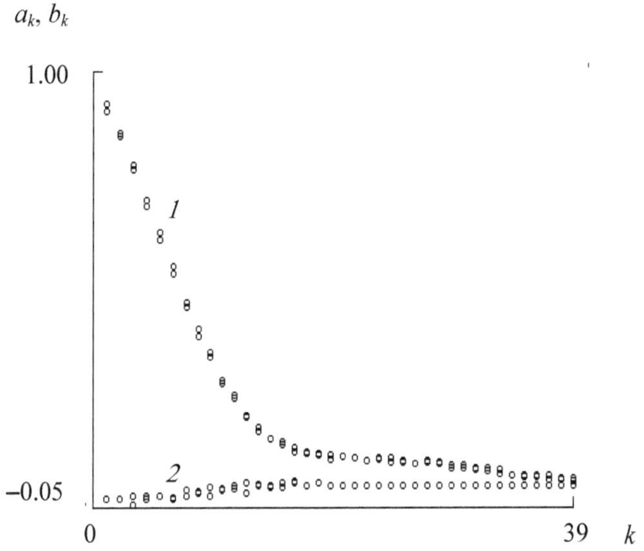

Fig. 9.4. Practical results of adaptive harmonic analysis of the diffraction line of the test specimen: (*1*) a_k; (*2*) b_k

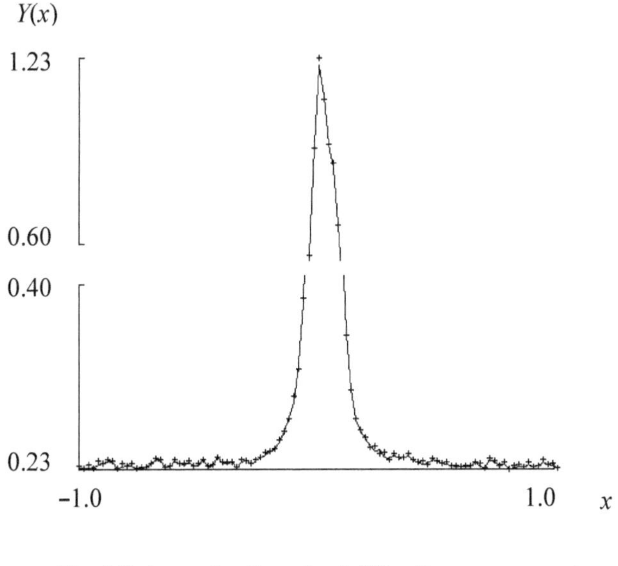

Fig. 9.5. Approximation of real diffraction measurements by Fourier coefficients self-correcting as to background

5. Requirements to the experiment on the accuracy of measuring the harmonics of the diffraction line. Study of inherent errors in the diffractometric data can specify the experimental conditions under which errors of determining harmonics of the diffraction line at least not become unallowable.

Essential time to measuring the diffraction line: Let the measuring interval be much larger than the width of the diffraction line, and the length of the Fourier series approximating the line profile be $K \to (M - 1)/2$ (M is the number of measurement points). Approximate covariance matrix of the trigonometric regression coefficients will consist of the elements $\left[MV_{\eta\eta}^{-1} + \omega k^2\right]^{-1} \delta_{kk}$ $(k \neq 0)$. The index η denotes the point number of the set $\{x_m\}$ such that $x_{m=\eta} = 0$.

When the number of points $M \gg 1$, the correction, depending on $\omega \sim K^{-2}$, is unimportant. Therefore, as a rough estimate of the variance of the measured harmonics c_k, when counting pulses at one point for a time t, one can take

$$\sigma_{c_k}^2 \approx \frac{1}{Mt} \frac{Y_0^{\max}}{\left[Y_0^{\max} - Y_0^{\min}\right]^2}.$$

Normalized harmonics $c_k' = c_k / c_0$ of high order k, when $c_k' \ll 1$, have variance $\sigma_{c_k'}^2 \approx \sigma_{c_k}^2 / c_0^2$. Here, $c_0 = \overline{y}$ as it follows from averaging over all m points of the variables of Eq. (9.3).

Introducing the limitation $\sigma_{c_k'}^2 < \varepsilon$ for the normalized harmonic of the highest order $k = K$, we obtain the formula for the indicative calculation of the total time of measuring the diffraction line that is $T = Mt$:

$$T \approx \frac{1}{\varepsilon} \frac{Y_0^{\max}}{\left[\overline{Y} - Y_0^{\min}\right]^2}.$$

Measuring time is determined with some margin. When the number of harmonics to be estimated is agreed to accuracy of data the controlled variance will be lower than established in calculation.

For specimen taken as a practical example, according to the preliminary data to measuring one point requires approximately 16 s when specified $\varepsilon = 10^{-4}$. By scanning in three passes with a total time of pulse count at point 15 s the variance of the normalized harmonic of largest order $K = 39$, at which achieved an acceptable approximation of primary measurements (Fig. 9.4–9.5), has an estimate of 0.16×10^{-4} whereas mean of the estimates 0.28×10^{-4}.

Formula derived to plan the time of the diffraction line measuring is useful for practice, especially when measuring a standard.

Allowed step of measuring the diffraction line: By measuring with a step β equal to the width of the pulse counter slit, instead of the true harmonics $\hat{c}_k$, distorted are appeared such as

$$c_k = \hat{c}_k \frac{\sin k\beta/2}{k\beta/2}.$$

Fourier coefficients of slit function decrease from unity to zero when $k \to (M-1)$, where M is the number of measurement points in the interval 2π.

It is naturally to require that the systematic distortion $\hat{c}_k - c_k$ does not exceed the limited standard deviation $\sigma_{c'_k} < \sqrt{\varepsilon}$, when $\sigma_{c'_k} = c'_k$ $(k = K)$.

Relative errors $(\hat{c}_k - c_k)/\hat{c}_k < 1$, if $\cos\frac{k\beta}{2} > 0$ (it is taken into account that $\cos x < \frac{\sin x}{x} < 1$ $(-\pi < x < \pi)$ [37]).

Hence, there is a condition under which the systematic error will not exceed the allowable limit:

$$K\beta < \pi.$$

Let us suppose that shape of the diffraction line is described by Lorentz function (similar Cauchy function) $\varphi(x) = \left[1 + \lambda^2 x^2\right]^{-1}$ with normalized harmonics $\hat{c}'_k = e^{-|k|/\lambda}$.

Equality $\hat{c}'_k = \sqrt{\varepsilon}$,

$$K = -\frac{\lambda}{2}\ln\varepsilon$$

satisfies, where $\lambda = \frac{\pi}{B}$, and B is the integral width of the function $\varphi(x)$.

So, scanning step is limited by the condition

$$\frac{\beta}{B} < -\frac{2}{\ln\varepsilon}.$$

If $\varepsilon = 10^{-4}$ specified, the allowable step is $\beta < B/5$.

Test specimen there presented was measured with step of approximately equal to $B/6$. Approximate integral line width is calculated as ratio of the integral measured intensity to the maximal:

$$B \cong \frac{2\pi}{M-1}\sum_{m=1}^{M} y_m \quad \left(y_1, y_M \text{ included with the weight } \tfrac{1}{2}\right).$$

In the same way, it can be determined that the allowable step for Gaussian shape of diffraction line is $\beta < B/3$.

§ 9.2. Determination of the Physical Profile of Diffraction Line with the Best Approximation to the True

Physical profile of the diffraction line containing information on the object under investigation is distorted by the "instrumental" convolution. Equation of convolution by applied to inaccurate experimental data becomes a structural model that able to be used for statistical estimation of harmonics of the physical profile.

1. Estimation of harmonics of the physical profile of the diffraction line by the method of pseudo-maximum likelihood. Fourier transform of the equation describing the observed shape of the diffraction line:

$$\varphi(z) = \int_{-\infty}^{\infty} \Phi(z-\zeta)\psi(\zeta)d\zeta,$$

where $\Phi(z)$ is the physical profile, and $\psi(z)$ is the instrumental profile, connects the vectors of the true harmonics of the functions entering into it:

$$\hat{\mathbf{c}} = \mathbf{L}\left(\hat{\mathbf{C}}\right)\hat{\mathbf{w}}. \qquad (9.5)$$

Here, **L** is matrix with the vector $\hat{\mathbf{C}}$ along its diagonal, the remaining elements of which are equal to zero.

In practice, the instrumental profile $\psi(z)$ is replaced for measured reference line. Usual calculation of harmonics of the physical profile is performed by the formula $C_k = c_k / w_k$. Here, c_k and w_k are estimates of Fourier coefficients $\varphi(z)$ and $\psi(z)$ from the experimental data.

With inaccurate (c_k, w_k) the deviations from the true harmonics $\hat{C}_k$ with probability P have the value $\left|C_k - \hat{C}_k\right| / \hat{C}_k \approx (\xi_P\sigma)/|\hat{c}_k|$, where σ^2 is the variance of the measurements, and ξ_P is the percent distribution points; therein is taken into account that for slowly decreasing $|\hat{w}_k| \gg \sigma$. When measured $\left|c_k\right| \approx \sigma$, hypothesis $\left|\hat{c}_k\right| = 0$ can not be excluded with the risk of $(1 - P) \leq 0.2$, even under the assumption of a normal distribution of errors. Such is the risk of arbitrarily large deviations of C_k from true values at any measurement accuracy.

The first approximation to an optimal estimation of the physical profile of the diffraction line was the regularization of the solution of the inverse problem of "instrumental" convolution according to A.N. Tikhonov [95]. To harmonize the regularization parameter with the errors of data it was required repeated experiment [35, 79].

However, if there are data of repeated measurements, special regularization becomes unnecessary. The problem is solved by methods of mathematical statistics, most appropriate to objective conditions.

Equation (9.5) in usual substitution of the data (**c**, **w**) as variables for even the true vector $\hat{\mathbf{C}}$ is satisfied only approximately. With reducing to the form

$$\mathbf{h} = \mathbf{c} - \mathbf{L}\left(\hat{\mathbf{C}}\right)\hat{\mathbf{w}}, \tag{9.6}$$

where **h** is a vector of inaccurate restrictions, it turns into a structural model of the data with unknown parameters $\hat{\mathbf{C}}$.

Let in a series of experiments be obtained certain vectors of Fourier representations: ($\mathbf{c}_1$, $\mathbf{c}_2$, ...) for the profile $\varphi(z)$ under study; ($\mathbf{w}_1$, $\mathbf{w}_2$, ...) for the instrumental profile $\psi(z)$. Then we can constitute a pair sampling that contains n elements ($\mathbf{c}_\mu$, $\mathbf{w}_\nu$) ($\mu = 1, 2, \ldots$ ($\nu = 1, 2, \ldots$)). The vector **C**, for which the likelihood function of a pair sampling is maximal, will be an estimate of the harmonics of the physical profile $\Phi(z)$.

Distribution of random variables **h** deviates from the normal one since there are distortions generated by constriction of infinite-dimensional $\left(\hat{\mathbf{c}}, \hat{\mathbf{w}}\right)$. The likelihood function is constructed as if these deviations would be absent. Maximization of the approximate function can bring in estimates with good statistical properties, although in reality the pseudo-maximal likelihood is obtained [2].

Let us introduce the indices $\lambda = \lambda(\mu, \nu)$ for pair observations $\left(\mathbf{c}_\mu, \mathbf{w}_\nu\right)_\lambda$. One pair observation of ($\lambda = \vartheta$) will be used for the initial estimate of $\mathbf{C}_\vartheta^0$ that is calculated from Eq. (9.5), as if the original data would be accurate. Over all other observations ($\lambda \neq \vartheta$) will being executed optimization. Turning over a pair sampling, we obtain a set of the estimates of $\mathbf{C}_\vartheta$ ($\vartheta = 1, \ldots, n$), and this will allow estimate their true error.

In order not to complicate the computational procedure, we suppose that vectors of inaccurate restrictions $\left(\mathbf{h}_1, \ldots, \mathbf{h}_n\right)$ are mutually independent and have one and the same, although unknown, covariance matrix of distribution.

Maximum likelihood estimation method gives the following objective function [2]:

$$\left.\begin{aligned} L(\mathbf{C}_\vartheta) &= \frac{n-1}{2}\log\det\mathbf{M}, \\ \mathbf{M} &= \sum_{\lambda\neq\vartheta}\mathbf{h}_\lambda(\mathbf{C}_\vartheta)\,\mathbf{h}_\lambda(\mathbf{C}_\vartheta)^{\mathrm{t}}. \end{aligned}\right\} \tag{9.7}$$

Problem is reduced to the search such vectors of $\mathbf{C}_\vartheta$ for which the ellipsoid of deviations from the convolution equation by the available pair sample is minimal.

Constraint vectors $\mathbf{h} = \{h_k\}$ are functions of the normalized harmonics precisely $\mathbf{c} = \{c_k\}$, $\mathbf{w} = \{w_k\}$ $(k = 1, \dots, K)$. The higher order of harmonics K is determined by a dimensionality of the vector $\{c_k\}$ sufficient for an adequate representation of the diffraction line of specimen. Correlation between the errors of harmonics of different orders is insignificant; therefore, the off-diagonal elements of the matrix $\mathbf{h}\mathbf{h}^{\mathrm{t}}$ can be neglected.

The solution of the formulated problem is a minimum point of the functional

$$F(\mathbf{C}_\vartheta) = \sum_{k=1}^{K}\left[\frac{n-1}{2}\log\sum_{\lambda\neq\vartheta}\left|h_k(C_{k\vartheta})\right|_\lambda^2 + \frac{1}{2}\omega k^2\left|C_{k\vartheta}\right|^2\right].$$

In addition to Eq. (9.7) hither with the coefficient $0 < \omega \ll 1$, the smoothing functional is included, which stabilizes the search results relative to fluctuations in the data sample [95].

The optimizing sequence of

$$\mathbf{C}_\vartheta^{i+1} = \mathbf{C}_\vartheta^{i} + \gamma(\Delta\mathbf{C}_\vartheta)^i \quad (i = 0, 1, \dots)$$

is constructed by the Gauss-Newton method. A step controlled by the coefficient $0 < \gamma \le 1$ is allowable if $F\left(\mathbf{C}_\vartheta^{i+1}\right) < F\left(\mathbf{C}_\vartheta^{i}\right)$.

Complex functions $h_k = \alpha_k - i\beta_k$ of the structural model at Eq. (9.6) depend on complex variables $c_k = a_k - ib_k$, $w_k = u_k - iv_k$ and complex parameters $C_k = A_k - iB_k$. For the iterative process of minimizing $F(\mathbf{C}_\vartheta)$, the following calculation formulas are obtained:

$$\Delta A_{k\vartheta} = -\frac{1}{d_k}\left\{\sum_{\lambda\neq\vartheta}\left[-\alpha_k u_k - \beta_k v_k\right]_\lambda + \omega k^2\left|A_{k\vartheta}\right|\right\},$$

$$\Delta B_{k\vartheta} = -\frac{1}{d_k}\left\{\sum_{\lambda\neq\vartheta}\left[\alpha_k v_k - \beta_k u_k\right]_\lambda + \omega k^2\left|B_{k\vartheta}\right|\right\},$$

$$\begin{cases}\alpha_k = a_k - \left[A_{k\vartheta}u_k - B_{k\vartheta}v_k\right], \\ \beta_k = b_k - \left[A_{k\vartheta}v_k + B_{k\vartheta}u_k\right],\end{cases} \quad d_k = \sum_{\lambda\neq\vartheta}\left|w_k\right|_\lambda^2 + \omega k^2.$$

Local minimum $F^* = F\left(\mathbf{C}_\vartheta^*\right)$ is achieved in one or two iterations with the initial $\omega_0 \sim 10^{-3}$. The way of dividing ω in half: $\omega_{j+1} = 0.5\,\omega_j$ ($j = 0, 1, \ldots$) leads to a neighborhood of the stationary point, where

$$\left|\frac{F_{j-1}^* - F_j^*}{F_j^*}\right| < \varepsilon, \quad \left|\frac{F_j^* - F_{j+1}^*}{F_{j+1}^*}\right| < \varepsilon \quad (\varepsilon \sim 0.01).$$

The point at which the magnitude of residual deviations $\mathbf{h}^* = \mathbf{h}\left(\mathbf{C}_\vartheta^*\right)$ is within their dispersion about the mean $\overline{\mathbf{h}^*}$, and therefore within the ranges of data errors, is considered to be the optimal estimate of $\mathbf{C}_\vartheta^*$, that is, the best of possible one.

The assumption on correspondence $\mathbf{C}_\vartheta^*$ to the measurement data is checked by means of a statistical criterion

$$\eta_\vartheta = \frac{\zeta}{n}\sum_{k=1}^{K}\sum_{\lambda=1}^{n}\left|h_{k\lambda}^*\right|^2 / s_{h_k}^2,$$

$$s_{h_k}^2 = \frac{1}{n-1}\sum_{\lambda=1}^{n}\left|h_{k\lambda}^* - \overline{h_k^*}\right|^2, \quad \overline{h_k^*} = \frac{1}{n}\sum_{\lambda=1}^{n} h_{k\lambda}\left(C_{k\vartheta}^*\right).$$

The correction $\zeta = 2(n-1)/(n-2)$ takes into account the presence of one adjustable parameter in the $(n - 1)$ equations of the structural model (9.5) and two variables with errors in each equation. The assumption will be rejected with reliability not less than P, if $\eta_\vartheta > K/(1-P)$ [2].

Computational expenses for the statistical estimation of harmonics of the physical profile are no more than to search for the optimal regularization parameter [95]. The procedure itself is simpler and more reliable. Over, when inexact instrumental profile, it is actual required to search for not one but two regularization parameters [32].

2. Tests of the method of determining the physical profile. An extent of approach of the estimates of $\mathbf{C}$ to the true vector of harmonics $\hat{\mathbf{C}}$ is checked with simulation experiments.

Let us take Lorentz function for the physical profile of $\Phi(z)$ and δ-shaped Gaussian function for the instrumental profile of $\psi(z)$. Fourier representations of the set out $\Phi(z)$ and $\psi(z)$, and of the output function $\varphi(z)$ are shown in Fig. 9.6.

Measured values of the Fourier coefficients $\{c_k\}$, $\{w_k\}$ are modeled by normally distributed random variables with mathematical expectation $\left(\hat{a}_k = \hat{c}_k,\ \hat{b}_k = 0\right)$, $\left(\hat{u}_k = \hat{w}_k,\ \hat{v}_k = 0\right)$ and with all the same prescribed variance σ^2. Higher order of harmonics K is limited by condition $\hat{c}_k \geq \sigma$.

In each series of simulation experiments, there are two minimally necessary measurements of c_k and w_k ($k = 1, \ldots, K$). The size of one pair-wise sampling of data $(\mathbf{c}, \mathbf{w})$ is $n = 4$. Total $M = 15\times10^3$ series of experiments was carried out. Estimates of the vector $\mathbf{C}$, obtained from the measurements, make up a sample of size $M\times n$, which with 100% probability contains not less than 99.95% of a distribution of general population [37].

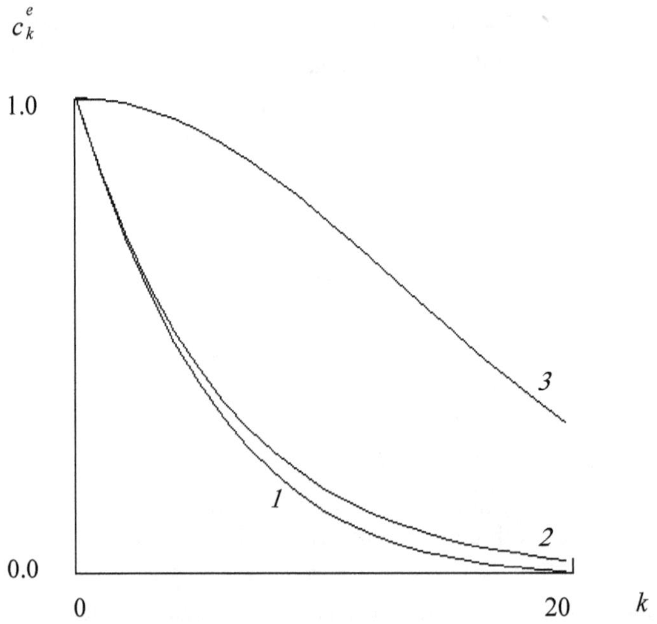

Fig. 9.6. Harmonics of model functions for simulation tests of the method of physical profile recovery: (*1*) $\varphi(z)$; (*2*) $\Phi(z)$; (*3*) $\psi(z)$

Figure 9.7 allows us to compare the true errors on the optimal estimation of the sought harmonics and by the traditional calculation. For simulation measurements of different accuracy, the growth curves of the relative errors $r_k = \hat{\sigma}_{C_k} / \hat{C}_k$ are constructed, where $\hat{\sigma}_{C_k}$ is the standard root mean square deviation of the estimates from the true harmonics:

$$\hat{\sigma}_{C_k} = \sqrt{\frac{1}{Mn} \sum_{m=1}^{M} \sum_{\vartheta=1}^{n} \left| C_{k\vartheta} - \hat{C}_k \right|_m^2} . \tag{9.8}$$

Errors of the initial approximations C_k^0 deriving from Eq. (9.5) are predicted theoretically. With known true values of the quantities $(\hat{c}_k, \hat{w}_k)$ and identical variance when measuring, the calculation formula takes the form

$$\hat{r}_k = \sigma \sqrt{\left[\hat{c}_k^2\right]^{-1} + \left[\hat{w}_k^2\right]^{-1}} .$$

An empirical curve r_k is approaching $\hat{r}_k$ with an improvement in accuracy of the data (Fig. 9.7).

In the sample of $C_{k\vartheta}$ ($\vartheta = 1, \ldots, n$) the errors are correlated. Correlation matrix varies little with the harmonics order k. There is no correlation between errors of harmonics C_k of different orders k.

The relative inefficiency of an estimate is generally measured by how much its generalized variance (covariance matrix determinant of a multidimensional random variable) exceeds the minimum possible [2].

One can calculate the inefficiency of the initial estimate $\mathbf{C}^0$ with respect to the optimal $\mathbf{C}^*$. For the diagonal covariance matrix of errors of the vector $\mathbf{C}$ having the dimension K, the following formula to the relative inefficiency is obtained:

$$e = \left[\prod_{k=1}^{K} \hat{\sigma}_{C_k^0}^2\right]^{1/K} \Big/ \left[\prod_{k=1}^{K} \hat{\sigma}_{C_k^*}^2\right]^{1/K} .$$

The participated here true variances $\hat{\sigma}^2$ of the elements C_k^0 and C_k^* are calculated by Eq. (9.8).

As it is uncovered, on different accuracy of the generated data, the relative inefficiency e of the traditional estimate of the sought harmonics in comparison with their optimal statistical estimate is approximately 1.9.

True standard error consists of a random dispersion of estimates relative to the distribution center, as the general mean $\langle C_k \rangle$, and the systematic bias of the center $\langle C_k \rangle - \hat{C}_k$:

$$\hat{\sigma}_{C_k} = \sqrt{\sigma_{C_k}^2 + \delta_{C_k}^2} .$$

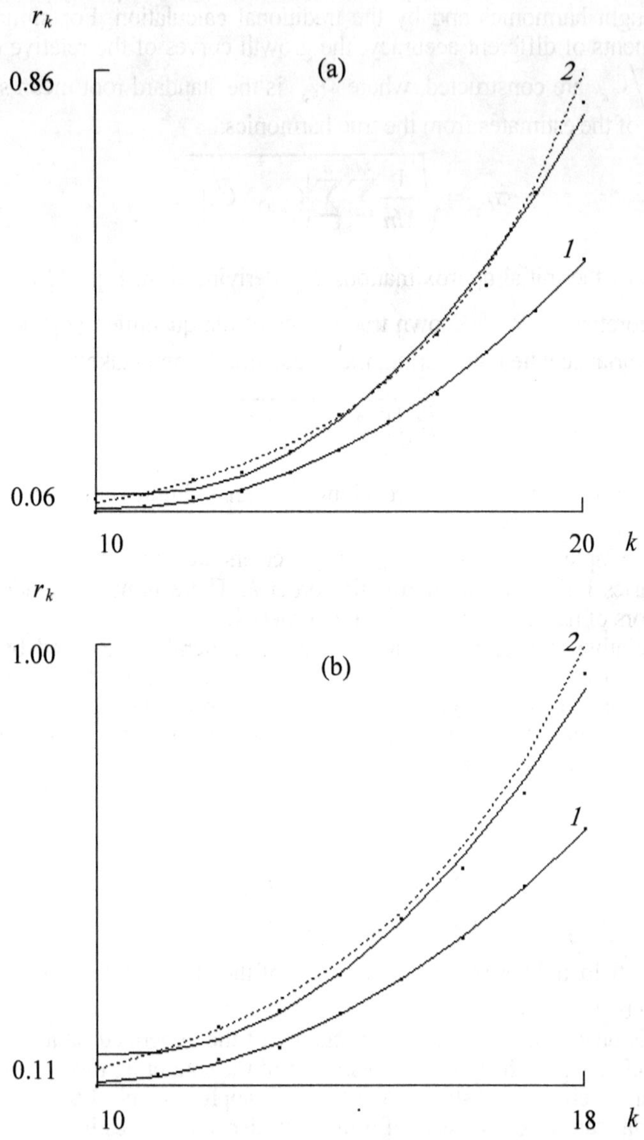

Fig. 9.7. Relative standard deviations from the true harmonics of the physical profile when data errors (a) $\sigma = 0.01$; (b) $\sigma = 0.02$, and method (*1*) statistical estimation; (*2*) traditional calculation. Dashed line indicates designed deviations

Here, δ_{C_k} denotes a bias of the estimates, increasing with the harmonics order of *k*.

In Table 9.4 provides information on the maximum relative biases of the estimates $\delta_{C_k} / \hat{C}_k$ calculated by different methods. For the traditional estimate C_k^0 the bias fraction in the true standard error $|\delta_{C_k}| / \hat{\sigma}_{C_k}$ is almost twice as large as for the optimal estimate C_k^*.

How much faster growing a random dispersion of the traditional estimates C_k^0, which close to distribution variance $\sigma_{C_k}^2$, it can be seen in Fig. 9.8.

Table 9.4.

The greatest systematic errors on reversing the convolution by data of different accuracy

Deviation measure	σ = 0.01 (*K* = 20)		σ = 0.02 (*K* = 18)	
	1	*2*	*1*	*2*
Relative bias to the true value	0.15	0.43	0.23	0.56
Bias portion in standard error	0.30	0.53	0.37	0.59

(*1*) statistical estimation; (*2*) traditional calculation.

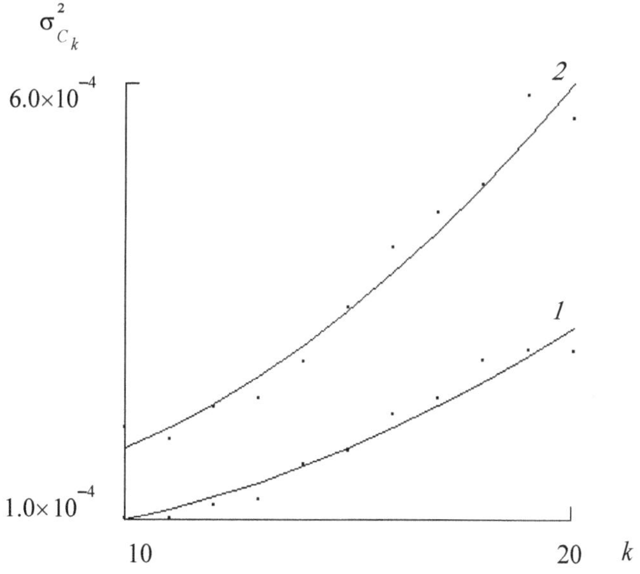

Fig. 9.8. Dispersion relatively the mean estimates of Fourier coefficients of the physical profile: (*1*) statistical estimation; (*2*) traditional calculation (σ = 0.01)

Based on the results of simulation tests, it can be concluded that the statistical method applied significantly improves the efficiency of estimating the vector of the sought harmonics, limiting both bias and variance.

Introduced optimization procedure allows without additional calculations to estimate the true error of estimates C_k^* by J. Tukey method known as "jackknife" [15].

An error of estimates obtained from n sub-samples of size $(n-1)$ with the successive exclusion of one element has the formula

$$\hat{s}_{C_k} = \sqrt{\sum_{\vartheta=1}^{n} \left| C_{k\vartheta} - \overline{C}_k \right|^2}.$$

Here, $\overline{C_k}$ is the average of the sample estimates $C_{k\vartheta}$ ($\vartheta = 1, \ldots, n$). To obtain an unbiased estimate of the true standard deviation, it is required multiply $\hat{s}_{C_k}$ by $\sqrt{n/(n-1)}$ [15].

In Table 9.5 the errors of the optimally estimated harmonics C_k by the simulated tests are summarized. The average sample estimate of the true standard error $\sqrt{\sum \hat{s}_{C_k}^2 / M}$ is biased down, despite the correction involved; because of small size of each re-sampling [15].

Probability distribution for estimates of C_k is unknown; therefore the significance of the obtained harmonics is tested using the Chebyshev inequality [37]. If $\left|\overline{C_k}\right| > \hat{s}_{C_k} / \sqrt{n(1-P)}$, then there will be $\left|\hat{C}_k\right| > 0$ with reliability of not less than P.

Under expensive real experiment apparently we will have to limit itself to the minimum required sample of data and take into account that the estimates of the true errors of obtained harmonics of the physical profile are understated.

Table 9.5.

Standard deviations of the estimates at the statistical method for reversing the convolution

Calculated values for harmonics C_k	*Errors in simulation measurements data*			
	$\sigma = 0.01$		$\sigma = 0.02$	
	$k = 1$	$k = 20$	$k = 1$	$k = 18$
Error by the mean of sample estimates, $\hat{s}_{C_k}$	0.0068	0.0152	0.0142	0.0238
Expected random error, σ_{C_k}	0.0092	0.0177	0.0187	0.0290
True error, $\hat{\sigma}_{C_k}$	0.0099	0.0185	0.0196	0.0313

Table 9.6.

Parameters of the dislocation structure of low-carbon steel specimens measured from harmonics of the diffraction line

Deformation extent	*Method for determining the physical profile*			
	statistical estimation		*traditional calculation*	
	Dislocation density [cm^{-2}]	*Fluctuations of density*	*Dislocation density* [cm^{-2}]	*Fluctuations of density*
18%	$[1.5; 1.8]\cdot 10^9$	[0.19; 0.25]	$[1.8; 2.2]\cdot 10^9$	[0.15; 0.20]
43%	$[2.8; 3.6]\cdot 10^9$	[0.11; 0.16]	$[2.3; 3.0]\cdot 10^9$	[0.14; 0.20]

3. Verifying the statistical estimate of the physical profile by practice. On effectiveness of optimization in applying to the real experiment it can be decided by the amount of reliable information retrieving from harmonics of the physical profile of the diffraction line.

Let us consider an example of diffraction analysis of a microinhomogeneous dislocation structure of low-carbon steel specimens with different extent of deformation. X-ray measuring was performed by D.A. Kozlov, as detailed in § 8.3.

Original data for determining the parameters of the structure presented in Table 9.6, these are the optimal harmonics of the physical profile of $\mathbf{C}^*$ and their initial approximations of $\mathbf{C}^0$ calculated by traditional way. In a real experiment it is useful to calculate $\mathbf{C}^0$ with the regularization coefficient $\alpha \sim 10^{-4}$ [79].

In view of the underestimation of the errors, the required reliability of harmonics significance is P = 0.999.

Approximate 90% confidence intervals for the average dislocation density and its relative fluctuations on variously oriented crystals are constructed using the automated system for studying the dislocation structure of polycrystals. In Table 9.6 the average intervals for 60 random samples of size 60 from the total sample of 900 measured values of parameters are given.

Loss of information under the traditional calculation is such that separation of the average dislocation densities in the test specimens has reduced on order, and the fluctuations of the densities became quite indistinguishable.

Practical results verified the good quality of the statistical estimate of the physical profile of the diffraction line.

CHAPTER 10
AUTOMATED SYSTEM OF RESEARCH OF DISLOCATION STRUCTURE OF POLYCRYSTALS

Diffraction studies of deformed polycrystals are supported by an interactive software system consisting of two subsystems; these are analysis of diffraction measurements and identification of dislocation structure models. The first subsystem realizes the optimal, by results accuracy, methods for the processing of diffractometric data. The identification subsystem provides experimentation with the models of diffraction observations for different structure states of the object [86].

§ 10.1. Studying the Dislocation Structure by Set of Models being Identified

Ensemble of the models for observations on the object shows up a measurable effect of the dislocation structure of a polycrystalline system on harmonics of the physical profile of the diffraction line:

$$A_k \cong F(x_k, \boldsymbol{\Theta}) + u_k \quad (k = 1, 2, \ldots).$$

The variable x_k includes the order of the harmonics k/æ, where æ is the fraction of the observation interval from the period in the diffraction space, the indices of the measured line $\{HKL\}$, and the crystallo-geomertric coefficient for existing slip systems. Systematic errors both the model itself and the data that increase with k are compensated by the auxiliary variable u_k (§ 7.1).

Object model parameters are enclosed at the vector $\boldsymbol{\Theta}$. Functions of the observations model $F(x_k, \boldsymbol{\Theta})$ are equipped with various methods of estimation of $\boldsymbol{\Theta}$ based on the maximum likelihood principle. The vector $\boldsymbol{\Theta}$, for which agreement of the observations model with the data $\{A_k\}$ by statistical criteria is not rejected, is considered to be a measured value of the vector of parameters of the dislocation structure. Searching the result of the measurement out of sampled start points $\boldsymbol{\Theta}_0$ uniformly distributed in the area of physical limitations is a computational experiment with the model of observations over the object.

1. Mathematical processing the data of diffraction measurements. Vectors of accumulated pulses under X-ray scanning the investigated profile of the line (specimen) as well as the instrumental profile (standard) to be excluding are inputted to the computer either from an external storage device or from a hard disk, by requested address under the program controlling. Two independent measurements of the specimen and standard are required. Reference information on the files structure of the original data is accompanied by a demonstration of an example.

Allowability of data and its quality are checked at the stage of preliminary processing. If the data is not rejected, the X-ray wavelength, and the diffraction line measurement interval as well as the type of crystal lattice (bcc, fcc) are requested to calculate the physical constants.

Automated system of research creates for itself the most accurate information support, realizing new methods of processing measurements; these are adaptive harmonic analysis of the observed diffraction line, furthermore optimal estimation of the diffraction line physical profile representing the object.

1. The adaptive method of harmonic analysis of the diffraction line realized the following principles: Fourier series is constructed from the coefficients self-correcting relative to the background present in the measurements; the length of the series approximating the line profile is consistent with the accuracy of primary data.

Sequential processing module performs an iterative search for the best self-adjusting regression model of observations and provides the Fourier coefficients of the analyzed line profile that are optimal in accuracy and stability. Real errors, including systematic deviations, are seen from the pair differences of the harmonics approximating the repeated measurements (§ 9.1).

The service program conducts a documented protocol of data processing in the made workbook and provides operative information on statistical conclusions from the analysis of the experiment. Examples of graphical display of the results of the adaptive harmonic analysis of measurements are shown in Fig. 9.4 – 9.5.

2. Harmonics of the physical profile of the diffraction line are estimated as parameters of an inaccurate structural model of the observed profile arising from the equation of "instrumental" convolution. Vector of harmonics of the physical profile is chosen so that discrepancies with the convolution formula are within the data errors.

Module of the joint processing generates a pair sampling from two measurements of the specimen and the standard and performs a search for as many independent stable estimates of the vector of harmonics of the physical profile as can be extracted from the existing pair sampling of data (§ 9.2).

In the protocol, the sample vectors of estimates are presented in the form of a table together with information on their covariance matrix. Dispersion of estimates is depicted graphically, for example, as in Fig. 8.4.

Allowability of the received harmonics is checked for the predicted second moment of the physical profile. With poor original data, it can turn out to be negative. According to acceptable estimates, the physical profile asymmetry coefficient and its sample variance are calculated.

Vectors of significant harmonics of the physical profile of the diffraction line form the core of information support for the subsystem of identification of dislocation structures. Significance of harmonics in view of systematic errors should have a confidence of not less than 99%. If the experiment did not provide reliable information, the program execution is terminated.

Researcher is provided with the means to manage the data organized by program. Unused data can either be destroyed, or moved to the archive.

2. Identification of the models for the plastic deformation structure of polycrystals. To determine the dislocation structure of plastic deformation, those harmonics A_k, are the most informative, in which the small value $(k/æ)/Q_{HKL}$ (§ 6.3). Therefore, it will be optimal measuring a diffraction line with large $\{HKL\}$ at the maximum possible interval. In the case of a sharp crystallographic texture, in order to correctly chouse $\{HKL\}$, it is required to know the parameters of the distribution of the crystal orientations, an example is considered in § 8.3.

Procedure for identifying a system of dislocations from the measured harmonics of the diffraction line includes: in the first stage, a nonparametric method; at the second stage, parametric methods that correspond to the circumstances.

1. Nonparametric identification allows test for accordance the chosen model from the provided list to the object under study. Method of nonparametric identification is a sequential linear regression analysis of the observations $Y_k = -\ln A_k$ over the object. Into the observations vector $\{Y_k\}$, initially having the smallest dimension, higher-order harmonics are sequentially included until a discrepancy $\{Y_k\}$ with the regression function, which is a second-order polynomial in k, is found.

For example, in Fig. 8.4 the regression curves are depicted, which were made by a successive analysis of harmonics of the diffraction line {112} of two specimens of low-carbon steel with different extent of deformation. In the specimen with 18% deformation, all reliably harmonics are fitted into one regression curve. When deformation of 72% the higher-order harmonics are deviated from the regression line of the lower harmonics, thereby the heterogeneity of the reflecting crystals is revealed.

The model of a mixed dislocation structure of a polycrystal will not be rejected only for the second specimen. The model of a microinhomogeneous (or homogeneous) dislocation structure is not discarded for both specimen. When in reality the dislocation structure is mixed, as in second specimen, but the model of a microinhomogeneous (or homogeneous) structure of a polycrystal is chosen, the analysis of the lower harmonics will show a “smoothed” structure approaching the main structure in specimen (§ 8.3).

If the discrepancy with the regression function under a sequential analysis of the observations $\{Y_k\}$ is detected in initial of k, the model of the plastic deformation structure is rejected in general. Here, either data involve serious errors, or structure existed is of another physical nature (§ 7.2).

2. Parametric identification of a complex dislocation structure of a polycrystalline system is performed on the stood out main parameters in models of different levels.

Parameters of a random system of dislocations inside crystals there are determined in the approximation of a homogeneous polycrystal. Fluctuations in distributions of sizes and coordinates of dislocation loops relative to the average distributions for a polycrystal are all the same indistinguishable. Neglect by fluctuations of the dislocation density with nonuniform deformation of the crystals receives to a bias in the estimates of its average value in a polycrystal $\overline{\rho_d}$.

Transferring the structural analysis from individual crystals to a polycrystalline system as a whole makes as the main parameters that the dislocation density distribution over differently oriented crystals. Details of the dislocation structure merge, turning into a unified parameter of defects in crystals. In the process of measuring the nonuniform density of dislocations in a polycrystal, the "over" parameters of the model of object are limited in the region of allowable values.

Microinhomogeneity of the dislocation density arising from a dispersion of the orientations of crystals is measured by the relative standard deviations of the average densities in equally oriented crystals from the total average for all crystals:

$$\chi = \sqrt{\overline{|\Delta\rho_d|^2}} \Big/ \overline{\rho_d}.$$

To identify the model of a homogeneous dislocation structure of a polycrystal, the harmonics of the diffraction line of low-order are sufficient. Information on the inhomogeneity is contained in the harmonics of the higher-order.

In the deformation texture, there are components with well-ordered crystal orientations. Fluctuations in the dislocation density $\overline{|\Delta\rho_d|^2}$ due the imperfect order still significantly exceed the Poisson variance, as it is revealed by analysis of thin-sheeted low-carbon steel (Table 8.6). But with the macroscopic inhomogeneity that exists, fluctuations $\overline{|\Delta\rho_d|^2}$ can be neglected. Main parameters become the average dislocation densities $\overline{\rho_d}$ in the components of a mixed structure of a polycrystal.

Weight fractions of dissimilar crystals involved in the reflection are predicted from the measured distribution of crystallographic orientations (§ 8.1). Without recourse to texture analysis, only it is known that weight fractions are in the interval (0, 1), where they will have to be sought when separately measuring the average dislocation densities in the structural components.

Each model of the object has its own software module that performs computational experiments and statistical analysis of their results. A series of experiments is planned to obtain sampling of measured values of the vector of structural parameters of size 900, keeping in mind that with probability of 99.9% between its extreme values it is located 99% of a population distribution [37].

Because the confidence intervals obtained from large samples are biased as a consequence of errors in the model, there is organized the extraction of a set of random samples of 60 measured values of the vector of parameters. This is a sufficient sample size, in order to construct a self-correcting along a sampling distribution the approximate confidence intervals of estimates such that the confidence probability of P is not less than the given probability (§ 7.1). An evolved average of the confidence intervals for parameters of the dislocation structure is recorded, for example, as in Table 8.4.

Method of interval analysis will allow to determine how the parameters of the structures under consideration are reliably distinguishable, assuming an equal probability of all values of the parameters in a range of variations of confidence intervals for estimates [93].

Sampling distributions of the measured values of parameters are represented in the form of histograms. An example showing the distribution of the measurements of inhomogeneous dislocation density is Fig. 8.7. By the histograms maximum are determined the most plausible estimates of the average dislocation density and its fluctuations over the crystal orientations.

With a macroscopic inhomogeneity of the dislocation density in the crystallographic texture, an image of the reproduced components of the intensity distribution of X-ray scattering is created, as in Fig. 8.6.

In the process of constructing and test of models, a versatile study of the inhomogeneous dislocation structure of deformed polycrystals is carried out.

3. Organization of studies for the structure of martensite transformation on the example of carbon steel. Method of constructing a model of dislocation structure arising by shear transformation of the crystal structure is adapted to the real phase state of metal systems [68].

Main parameters of the phase and dislocation structure of martensite formed during quenching of steel are determined. Chosen model agrees with the experiment that proven the existence of a system of small dislocation loops in martensite (§ 7.2). Original data are the harmonics of the physical profile of the diffraction line, measured with all the best accuracy (§ 9.1 and § 9.2).

Investigations were carried out on quenched specimen of U10 steel. As a reference representing an instrumental profile, an annealed specimen of steel 01Yu, where the concentration of carbon in 100 times less, is used.

The angles intervals of X-ray measuring of the diffraction lines in Co-K_α radiation are follows

$$\Delta 2\theta_{\mathtt{W-B}} = \begin{cases} 47.8^\circ - 57.0^\circ & \{110\}, \\ 94.8^\circ - 104.8^\circ & \{112\}. \end{cases}$$

The scanning step is 0.1° with the pulse counting time of 10 s per point. Number of passes for the specimen is 3, and for the reference 6.

Process is repeated with a new installation of the specimen and reference, as prescribed by the method of optimal estimation of the physical profile of the diffraction line (§ 9.2). X-ray measuring has been performed by D.A. Kozlov.

1. Estimation of the diffraction spectrum model. In the phenomenological model of the constitution of the arising spectrum of X-ray lines, all available information on the martensite phase state is represented. The optimal by accuracy and stability method of the spectrum model estimation was developed in [69–70].

In the iron-carbon system, a two-phase martensite is revealed, consisting of crystals with different tetragonality extent of the body-centered lattice. A diffraction spectrum of two-phase martensite with an impurity of the retained austenite that is matrix γ–phase is under consideration.

Equations of the model describe the coefficients of Fourier representation of the multiplet observing in the experiment:

$$\mathcal{A}_k = A_k \sum_i w_i \cos(kz_i), \quad \mathcal{B}_k = A_k \sum_i w_i \sin(kz_i).$$

Main multiplet {110} consists of five components with coordinates $z_0 < z_1 < z_2 < z_3 < z_4$ in the normalized interval of observations (–1, 1). The coordinate z_0 refers to the {111} line of fcc crystals of the γ–phase.

Weight fractions of the martensite components are bound by the restrictions[1]

$$\begin{cases} \frac{1}{3}(1-w_0) < w_1 < \frac{2}{3}(1-w_0), \\ w_2 = \frac{2}{3}(1-w_0) - w_1, \\ w_3 = \frac{1}{2}w_2, \quad w_4 = \frac{1}{2}w_1; \end{cases}$$

the weight fraction of γ–phase according to the measurement data on the amount of retained austenite in carbon steel is in the range ($0 < w_0 < 0.2$) [52].

In the model of {112} multiplet, required to check the reliability of the revealed phase structure of martensite, the rearrangements are occur: $w_1 \rightleftarrows w_4, w_2 \rightleftarrows w_3 \left(w_0 = 0\right)$.

The number of independent parameters of n-component model, including the vector of being estimated harmonics $\{A_k\}$ (k = 1, 2, … , k_{max}) of the approximate shape of spectral lines, is equal to $k_{max} + 2n - 3$.

---

[1] The effect of weakening of the intensity of diffraction due to displacements of the lattice sites around the interstitial carbon atoms for doublets is insignificant [70].

In an initial approximation, harmonics A_k ($k > 1$) decrease as $\sim k^2$; the first-order harmonic is modeled by a random variable close to unity. With a random choice of starting points (**w**, **z**), which satisfy all restrictions, there is searched the global minimum of the objective function of the method [69].

Systematic errors of the model and data, which according to statistical checks are comparable to random dispersion, are compensated for the centering of deviations from the model relative to their overall average. The number of degrees of freedom is reduced and, in order to degrees of freedom remain for check the adequacy of the model, it is required $k_{max} > 2n - 2$.

Optimum is considered achieved if the agreement of the predicted and observed spectrum is not rejected by a statistical check, and moreover the estimates of all parameters are significant and not correlated. A smoothest curve for approximation of the profile within the data dispersion is selected. Stabilized in this way solution produces the least biased estimates [69–70].

Figure 10.1 depicts the diffraction spectra of martensite of carbon steel with separated lines. The lines profile was restored from the measured harmonics using the method of stable summation of Fourier series [95].

In Table 10.1 are given the sample mean and standard deviations of estimates of the phase structure of martensite (sample size of 17). To calculate the carbon content in phases α and γ, data are used on dependence of the lattice periods by the concentration of the interstitial solid solutions [60].

There is not detected the presence of significant discrepancy in the information on the phase state of martensite of carbon steel provided by different multiplets.

Table 10.1.

Estimates of the phase structure parameters of the carbon steel martensite

Phase components	*Weight fraction*	*Tetragonality extent*	*Carbon concentration [wt %]*
α – high-carbon	0.5620 ± 0.0227	1.0331 ± 0.0010	0.7022 ± 0.0222
α – low-carbon	0.2638 ± 0.0130	1.0065 ± 0.0033	0.1386 ± 0.0694
γ – austenite state	0.1742 ± 0.0134	—	0.3338 ± 0.0422
α – high-carbon	0.6789 ± 0.0143	1.0382 ± 0.0006	0.8107 ± 0.0122
α – low-carbon	0.3211 ± 0.0143	1.0074 ± 0.0022	0.1562 ± 0.0466

In the second row, estimates from {112} line measuring; before the subsequent {110} line measuring, the aging of martensite has occurred by precipitation of carbon with the formation of clusters [21].

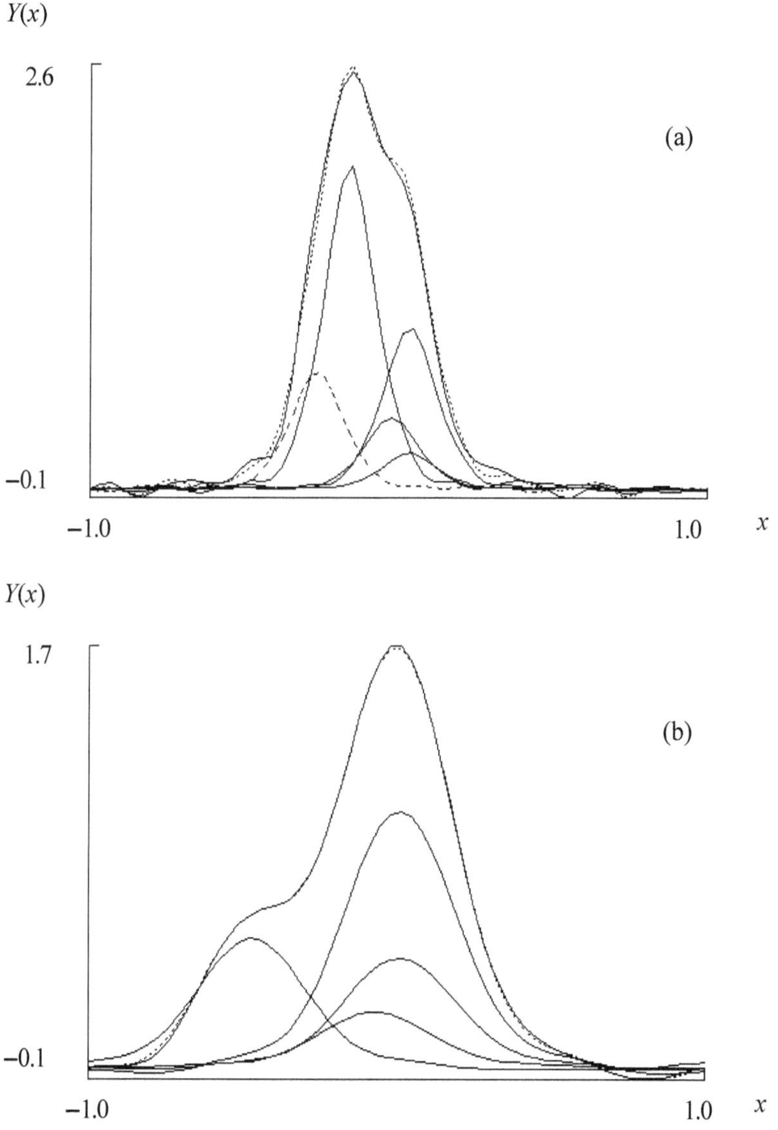

Fig. 10.1. Identified components of the diffraction multiplets of carbon steel: (a) {110}; (b) {112}. Dotted line shows the predicted profile of multiplet. Dashes mark off the {111} line of γ–phase

2. Estimation of the dislocation structure model. Theoretical model of observations describes the shape of the lines of the diffraction spectrum, based on conceptions of a system of dislocations in crystals. Having harmonics of the approximate shape of the spectrum lines, we will carry out a test of the model of martensite transformation structure.

Size of sample $\{\mathbf{A}_r\}$ ($r = 1, 2, \dots, m$) should provide a good covariance matrix of observations $\mathbf{V_A}$. Estimates of the distribution variance of σ_k^2 with samplings of small size are underestimated, so an acceptable sample size of $m \geq 10$.

Data there should be sufficient to estimate all the parameters of the diffraction observations model, including the approximation coefficients of systematic deviations by Legendre polynomials of a suitable degree l, so one must have $M \geq 3 + (l_{max} + 1)$ harmonics A_k, whose order k is not lower than the applicability limit onto the model Eq. (7.6).

Function $f_k(\mathbf{\Theta})$ of the model Eq. (7.6), specified on the interval $k \geq k_{min}$, is reduced to the linear regression function of the logarithms of (unbiased) harmonics $Y_k = -\ln A_k$. Due this, the lower limit of the interval k_{min} is found by the method of regression analysis of observations with successive exception of the lower-order harmonics ($k = 1, 2, \dots$) (§ 7.1).

For inhomogeneous martensite, only an inaccurate regression model of observations can be constructed. It is expected that the completely unknown systematic distortions present in the observations vector of $\mathbf{Y}$ become much smaller than Y_k at $k \to K$ due to more rapid decreasing of the higher-order harmonics of the polluting γ–phase. Therein the search for a spacing where deviations of Y_k from the regression equation are comparable with their errors leads to the allowed limit of k_{min}, although estimates of the regression coefficients are heavily biased.

Sequential regression analysis of the available data, actually, performed a nonparametric identification of the system of small dislocation loops in tetragonal crystals of martensite. Dispersion of observations data relative to the regression line can be seen in Fig. 10.2.

Further there is search for parameters using harmonics A_k belonging to the interval of observability of the system of dislocations on which $k \geq k_{min}$. Search method is to solve the optimization problem in the form of (7.7).

Dislocation density is measured with limitation of the remaining parameters of the dislocation structure model in the allowed intervals established from theory and experiment (§ 6.2 and § 7.2):

$$\begin{cases} 10^8\ \text{cm}^{-2} < \overline{\rho_d} < 10^{12}\ \text{cm}^{-2}; \\ 10 < \bar{\xi}/a < 30, \quad 0 < \sigma_\xi/\bar{\xi} < 0.3; \\ 0 < \bar{\tau}/b < 3, \quad 0 < \bar{\mu}/d < 3. \end{cases}$$

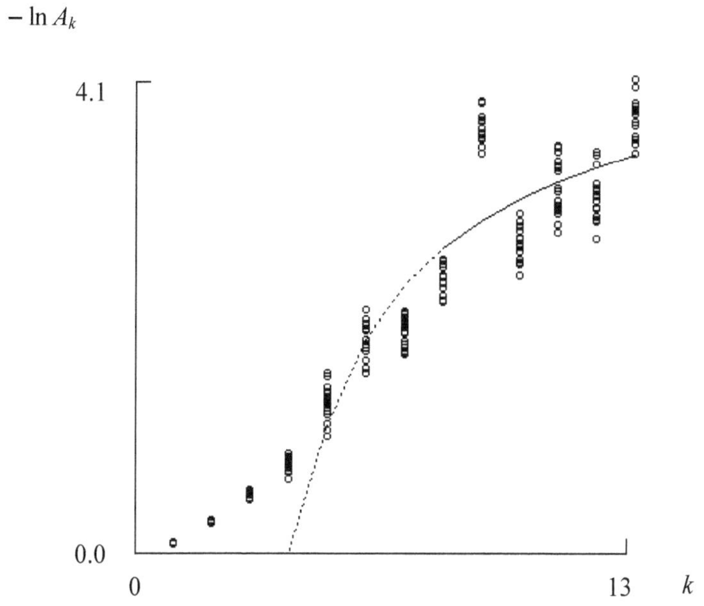

Fig. 10.2. Dispersion of harmonics of an approximate shape of lines in the model of {110} martensite multiplet; regression curve marks the interval of observability of the dislocations system (æ = 0.2295)

In a priori intervals, all parameter values are equally probable.

A good approximation of systematic errors at sufficient sample of original data is achieved by Legendre polynomials not above than the first degree ($l_{max} = 1$).

Tests conducted, with a confidence probability of not less than 90%, revealed for the dislocation structure of martensite transformation a limitation of the average radius of the loops $\bar{\xi}/a < 25$.

In Table 10.2 the estimates of the average dislocation density in martensite of carbon steel obtained from a series of computational experiments are presented.

Table 10.2.

The average density of dislocations in the martensite transformation structure

Specimen	*Approximate 90% confidence intervals of estimates*	*Most plausible estimates by the maxima of histograms*
U10	$[0.9; 1.0]\times10^{12}$ cm^{-2}	$(0.6–1.0)\times10^{12}$ cm^{-2}
01Kh5	$[0.7; 0.8]\times10^{11}$ cm^{-2}	$(0.6–1.0)\times10^{11}$ cm^{-2}

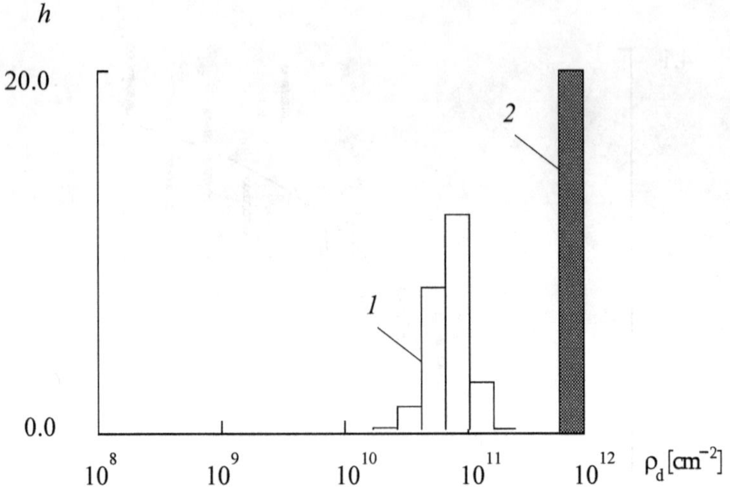

Fig. 10.3. Sampling distribution of the dislocation density measurements in the martensite crystals of steel: (*1*) 01Kh5; (*2*) U10.

Confidence intervals were produced an average of 900 random samples of size 60, extracted from the aggregate sample of measurements data 9×10^3.

To compare there are given the measurements data for dislocation density in a single-phase cubic martensite of steel 01Kh5 with carbon concentration not more than 0.01 wt % (§ 7.2 and § 7.3).

Sampling distributions of the measured values of dislocation density in the specimens under study is represented by histograms in Fig. 10.3.

In tetragonal martensite, the measured dislocation density became an order of magnitude higher than in cubic.

All available information on the dislocation structure of the martensitic transformation contained in the diffraction experiment can, in principle, be extracted using the data of analysis of two multiplets: {110} and {112}[1].

Figure 10.4 shows dispersion of harmonics of the profile of lines in the diffraction multiplet {112} with the identified components parameters presented in Table 10.1. A sample size is $m = 20$.

---

[1] When microtwins occur, the diffraction line {112} is inapplicable for dislocation structure analysis. The model (7.5) is rejected as inadequate to observations, for the parting of dislocations in twinning significantly affects the beginning harmonics.

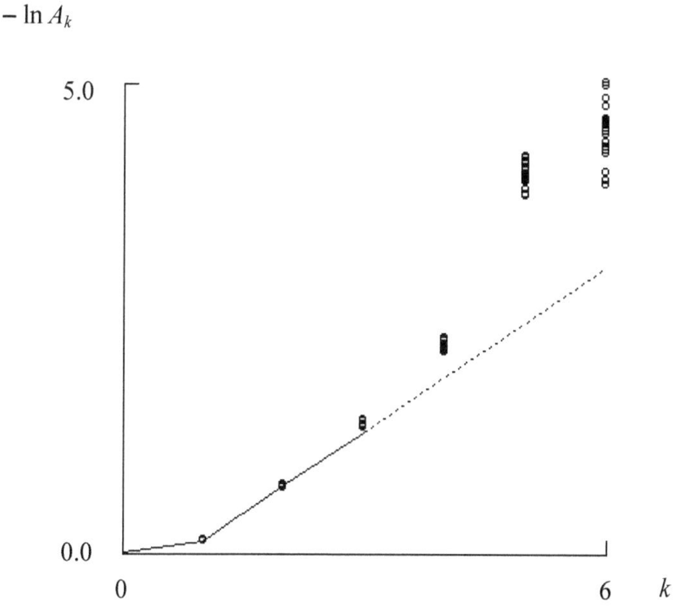

Fig. 10.4. Dispersion of harmonics of an approximate shape of lines in the model of {112} martensite multiplet; regression curve marks the interval of observability of long-range order in the dislocations system (æ = 0.1787)

Harmonics $\{A_k\}$ of the order $k \leq 3$ are statistically significant with a confidence of 99.9% under estimating the true variance $\sigma^2_{A_k}$ by the method "jackknife" [15]. The variance homogeneity is not rejected by statistical test [31].

Available data is applicable to determine the long-range order in the arrangement of dislocation loops on parallel slip planes.

The method of measuring dislocation density and order parameters by beginning harmonics of the diffraction line with large indices $\{HKL\}$ is presented in § 7.3. The transformation of the diffraction observations model for tetragonal crystals is described in § 7.4.

In Figure 10.5 the distribution of the measured values of the basic parameters of the dislocation structure model is presented while accumulated data sample of size 9×10^3.

Table 10.3 presents the average of 900 confidence intervals for being measured parameters that constructed over random samples of size 60, extracted from the total sample of measurements.

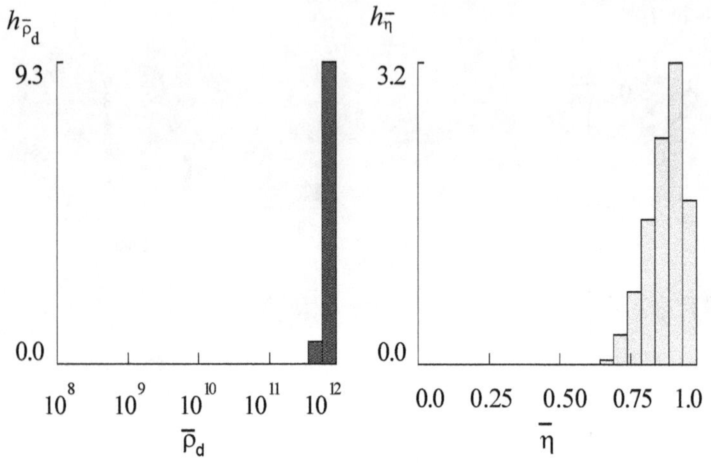

Fig. 10.5. Sampling distribution of the measurements of dislocation density – $\bar{\rho}_d$ [cm^{-2}] and of degree of order – $\bar{\eta}$ in the martensitic transformation structure of carbon steel

Table 10.3.

Long-range order in the dislocation structure of the martensitic transformation

Parameter measured	*Approximate 90% confidence intervals of estimates*	*Most plausible estimates by the maxima of histograms*
Dislocations density – $\bar{\rho}_d$	[0. 71; 0.73]×10^{12} см$^{-2}$	(0.6–1.0)×10^{12} см$^{-2}$
Degree of the order – $\bar{\eta}$	[0.88; 0.91]	(0.90–0.95)

Lower confidence limit of the order period is $\bar{\ell}/\bar{\xi} \gtrsim 1.0$.

The dislocation density estimate close to that expected according to analysis of the diffraction multiplet {110}.

By practical example it is reliably revealed the forming a periodic dislocation structure under the martensitic transformation of crystals.

§ 10.2. Simulation of a Random System of Dislocations on Measured Parameters

Uniform dislocation distribution in the slip plane is unstable. This is evidenced with the behavior of direct parallel dislocations, that prescribed by the equations of their moving in the stress field [26].

There are an infinite number of allowable microstates of a system of dislocations with the observed coordinates and sizes of loops at available their average concentration in crystal. Statistical modeling predicts the most probable microstate of a system of dislocation loops, which can be stable.

If there are many independent flows of loops from different activated sources, the summary flow in the limit can become Poisson one [59].

Simulated stream is a sequence of random vectors $\mathbf{r}_j$ ($j = 1, 2, \ldots$) that describe the coordinates of the centers of dislocation loops in the slip plane with equiprobable directions in the interval of angles $(0, 2\pi)$. Distances $r = |\mathbf{r}|$ between loops arriving one after the other are subject to the exponential distribution law:

$$\varphi(r) = \lambda e^{-\lambda r} \quad (\lambda > 0).$$

The most likely is the smallest distance r between the centers of loops. Average distance is the inverse value of the flux density: $\langle r \rangle = \lambda^{-1}$ [37].

Domain of setting of the Poisson flow can be divided into independent subregions [3]. For an allocated volume in crystal with the known average concentration of loops there is calculated the expected number of loops n, and the flux density λ is calculated from it. Realization of a random sequence forming a stream is calculated by the formula

$$r = -\lambda^{-1} \log z,$$

where z is a uniformly distributed random variable on the interval [0, 1] [17].

It is assumed that dislocation loops with equal probability can appear in all q types of slip planes in a crystal.

A random number of loops l that are in the ν-th pack of slip planes ($\nu = 1, \ldots, q$) under consideration, when their total amount in allocated volume of crystal of n, is determined from the probability experiments by the Bernoulli scheme [37]:

$$l = \sum_{i=1}^{n} \beta_i; \quad \beta_i = \begin{cases} 1 & \text{with probability } q^{-1}, \\ 0 & \text{with probability } \left(1 - q^{-1}\right). \end{cases}$$

The most probable value of the sum of n Bernoulli random variables is $(n+1)/q$ ($q = 6$ for bcc; $q = 4$ for fcc).

Random fluctuations of sizes appearing loops are considered to be approximately logarithmically normal with mean parameters in a polycrystal.

For the identified structure of low-carbon steel deformed by 32% using statistical modeling a graphic image of the distribution of dislocation loops in a pack of slip parallel planes around 0.5 μm in thickness was obtained. Between the acting slip planes a distance is accepted of ~ 200 Å.

Calculations were performed on the parameters of dislocations system randomly chosen from the confidence intervals of the estimates, where all their values are equally probable:

Dislocations density $\overline{\rho_d}$	*Average radius of loops* $\overline{\xi}/a$	*Variation coefficient* $\sigma_\xi/\overline{\xi}$	*Correlation radius* $\overline{\tau}/b$
$[1.8; 2.3]\cdot 10^9$ cm^{-2}	[460; 552]	[0.18; 0.22]	[1.5; 1.8]

Generated Poisson flow of loops with densities corresponding to the parameters of the structure of specimen revealed the frequent occurrence of stretched random accumulations of loops randomly deviding the pure regions of a crystal.

Figure 10.6 exhibited randomly originated forms of boundaries of multiply connected regions, which are inherent to the experimentally observed a cellular dislocation structure of deformed crystals.

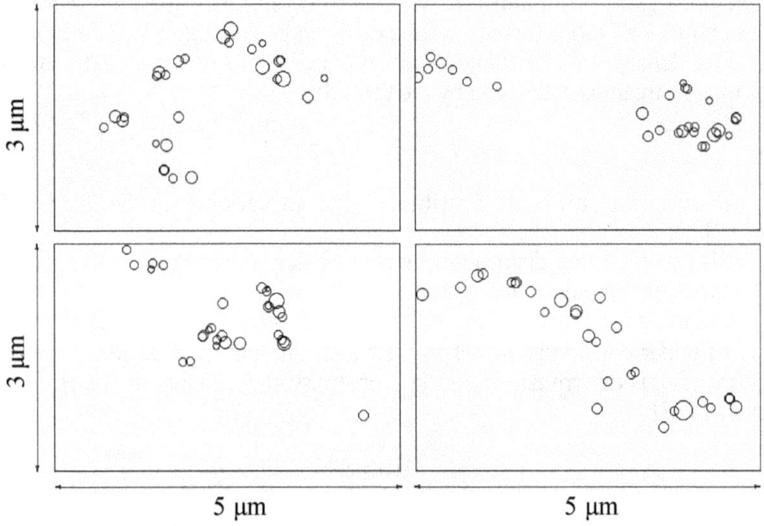

Fig. 10.6. Random accumulations of dislocation loops in a slip band after deformation of 32%; statistical modeling with measured parameters of structure of low-carbon steel

Visible microstates in the ensemble of random systems of dislocations coincide with experience. This leads to the conclusion that a stable system, generated from a large number of loops of dislocations with self-regulating randomness, is subject to a general description by the Poisson model.

§ 10.3. Conclusion on Practical Tests of the Dislocation Structure Studying Method

With plastic deformation of polycrystals, the average dislocation density therein by the estimates obtained increases within one order of magnitude, which is comparable to the observed increase in the yield strength as a result of strain hardening. With increasing deformation, the nonuniformity of the dislocation density distribution across the crystals decreases. Fluctuations of density, caused by dispersion of crystal orientations, lower regularly (Fig. 10.7).

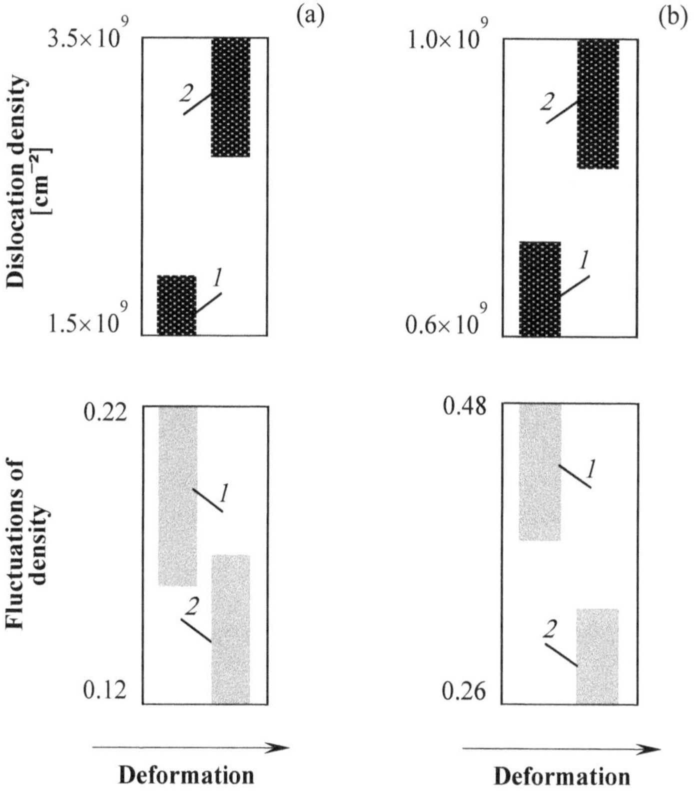

Fig. 10.7. Changing microinhomogeneous dislocation density with increasing extent of deformation of metals: (a) low-carbon steel after (*1*) 18% and (*2*) 43% deformation by rolling; (b) pure aluminum after (*1*) 20% and (*2*) 52% deformation by upsetting (data by D.A. Kozlov)

Various dislocation models of strain hardening of crystals derive to the same equation of the relation between the strength for plastic flow σ_0 and the dislocation density $\overline{\rho_d}$ as by original Taylor theory: $\sigma_0 \sim \sqrt{\overline{\rho_d}}$ [55]. Consequently, the measured relative fluctuations in the density of dislocations reveal a degree of inhomogeneity of strain hardening of a polycrystal, since approximately a ratio is true

$$\frac{\Delta\sigma_0}{\sigma_0} \approx \frac{1}{2}\frac{\Delta\rho_d}{\rho_d}.$$

A giant increase in the density of dislocations is related with sheared phase transformations in polycrystalline systems, whose picture is guessed from the structures that have arisen.

The texture transformation appears as a macroscopic shear, localized in an ensemble of crystals undergoing a breakthrough of constrained plastic flow (§ 8.3).

The martensitic transformation is accomplished by microscopic shears, jointly reassembling the crystal lattice (§ 7.2). In crystals with the displacements of reduced symmetry, the shears are multiplied, and the dislocation density reaches an ultimate value (§10.1).

Revealed periodicity in the dislocation structure of the martensitic transformation of crystals uncovers the wave nature of the process.

The excitation of long-wave lattice vibrations is theoretically explained by quantum transitions of "hot" electrons to lower energy levels on a sharp fall in temperature. It is known the process of emission of long-wavelength phonons (lattice vibration quants) by conduction electrons during low-temperature relaxation of the crystal system [46].

If the observed period of order ℓ is commensurate with the phonon wavelength, then the dislocation loops of a comparable radius $\xi \sim \ell$ (§10.1) dissipate phonons nucleated them [94].

Information on the states of the dislocation structure of polycrystalline systems, obtained in practice, is consistent with the physical representations, which proves the correctness of the theoretical propositions at the basis of the method.

REFERENCES

1. Barabash R. and Klimanek P., Phenomenological and microscopical description of scattering on different dislocation arrangements, *Z. Metallkunde* **92**, 70–75 (2001).
2. Bard Y., *Nonlinear Parameter Estimation* (Academic Press, New York, 1974; Statistika, Moscow, 1979).
3. Bol'shakov I. A. and Rakoshits V. S., *Applied Theory of Random Flows* (Sovetskoe radio, Moscow, 1978) [in Russian].
4. Borodkina M. M. and Spektor E. N., *X-ray analysis of the texture of metals and alloys* (Metallurgiya, Moscow, 1981) [in Russian].
5. Box J. E. P., *Robustness in the strategy of scientific model building. Robustness in statistics*, pp. 201–236 (Academic Press, New York, 1979; Mashinostroenie, Moscow, 1984).
6. Breitenberger E., Analogues of the normal distribution on the circle and sphere, *Biometrika* **50**, 81–88 (1963).
7. Bunge H.–J., *Mathematische Methoden der Texturanalyse* (Akademie Verlag, Berlin, 1969); *Texture Analysis in Materials Science. Mathematical Methods* (Butterworths Publ., London, 1982).
8. Clément A. and Coulomb P., Eulerian simulation of deformation textures, *Script. Metallurg.* **13**, 899–901 (1979).
9. *Computers, models, computational experiment*, Ed. A. A. Samarskii (Nauka, Moscow, 1988) [in Russian].
10. Cowley J. M., *Diffraction Physics* (Elsevier, New York, 1975; Mir, Moscow, 1979).
11. Cox D. R., Hinkley D. V., *Problems and Solutions in Theoretical statistics* (Wiley, New York, 1978; Mir, Moscow, 1981).
12. Davies G. J., Goodwill D. J. and Kallend J. S., Charts for analysing crystallite distribution function plots for cubic materials, *J. Appl. Cryst.* **4**, 67–70 (1971).
13. Draper N. R. and Smith H., *Applied Regression Analysis* (Wiley, New York, 1981; Finansy i Statistika, Moscow, 1987).
14. Ebeling W., *Formation of Structures in Irreversible Processes. Introduction to the Theory of Dissipative Structures* (Mir, Moscow, 1979).
15. Efron B., *Non-traditional Methods of Multivariate Statistical Analysis*, Collection of articles (Finansy i Statistika, Moscow, 1988).
16. Elliott J. P. and Dawber P. G., *Symmetry in Physics*, Vol. 1 and 2 (Macmillan Press, London, 1979; Mir, Moscow, 1983).
17. Ermakov S. M., *The Monte Carlo Method and Related Questions* (Nauka, Moscow, 1975) [in Russian].
18. Eshelby J. D., *The Continuum Theory of Dislocations*, Collection of articles (In. Lit., Moscow, 1963).
19. Fisher R., Dispersion on a Sphere, *Proc. Roy. Soc.* **A217**, 295–305 (1953).

20. Fridman Ya. B., *Mechanical Properties of Metals*, Vol. 1: *Deformation and Failure*, 3nd ed. (Mashinostroenie, Moscow, 1974) [in Russian].
21. Genin J. M. R. and Flinn P. A., Mössbauer effect study of the clustering of carbon atoms during the room temperature aging of iron carbon martensite, *Trans. Metallurg. Soc. AJME* **242**, 1419–1430 (1968).
22. Glansdorff P. and Prigogine I., *Thermodynamic Theory of Structure, Stability and Fluctuations* (Wiley, New York, 1971; Mir, Moscow, 1973).
23. Gnesin B. A. and Yashnikov V. P., The role of a deviation of primary beam from the goniometer horizontal plane in three-dimensional texture analysis, *Zavod. Lab.* **53** (3), 38–41 (1987).
24. *Handbook of Mathematical Functions*, Ed. M. Abramowitz and I. A. Stegan (Dover, New York, 1965; Nauka, Moscow, 1979).
25. *Handbook on Probability Theory and Mathematical Statistics*, Ed. V. S. Korolyuk (Nauka, Moscow, 1985) [in Russian].
26. Head A. K., A continuum model for two-dimensional dislocation distributions, *Philosophical Magazine A* **55**, 617–629 (1987).
27. Hill R., *Mathematical Theory of Plasticity* (Gostekhizdat, Moscow–Leningrad, 1956).
28. Hill R., Continuum micro-mechanics of elastoplastic polycrystals, *J. Mech. Phys. Solids* **13**, 89–101 (1965).
29. Jaakko Kajamaa, Determination of cold rolling and recrystallization texture in copper sheet by neutron diffraction, *Trans. Metallurg. Soc. AJME* **242**, 973–977 (1968).
30. Jenkins G. and Watts D., *Spectral Analysis and its Applications* (Mir, Moscow, 1972).
31. Johnson N. L. and Leone F. C., *Statistics and Experimental Design in Engineering and the Physical Science* (Wiley, New York, 1977; Mir, Moscow, 1980).
32. Karmanov V. G., *Mathematical Programming* (Nauka, Moscow, 1986) [in Russian].
33. Khayutin S. G., Anisotropy of plasticity of textured metal tape, *Tsvetnye metally* № 3, 80–83 (1983).
34. Kheyker D. M. and Zevin L. S., *X-ray Diffractometry* (Fizmatgiz, Moscow, 1963) [in Russian].
35. Kochetov I. I., On a new method for choosing the regularization parameter, *Zh. Vychislit. Math. and Math. Fiz.* **16**, 499–503 (1976).
36. Korin B. P., On the Distribution of a statistic used for testing a covariance matrix, *Biometrika* **55**, 171–178 (1968).
37. Korn G. A. and Korn T. M., *Mathematical Handbook for Scientists and Engineers: Definitions, Theorems, and Formulas for Reference and Review* (McGraw-Hill, New York, 1968; Nauka, Moscow, 1974).
38. Kozlov D. A., Petrunenkov A. A., Satdarova F. F. and Kekalo A. I., Texture of a thin sheet of low-carbon steel upon recrystallization under rapid heating conditions, *Izv. Akad. Nauk SSSR, Met.* №.5, 137–141 (1986).

39. Krivoglaz M. A., *Theory of X-ray and Thermal Neutron Scattering by Real Crystals* (Nauka, Moscow, 1967; Plenum, New York, 1969).
40. Krivoglaz M. A., Martynenko O. V. and Ryaboshapka K. P., Effect of correlation in arrangement of dislocations on X-ray diffraction by deformed crystals, *Fiz. Met. Metalloved.* **55**, 5–17 (1983).
41. Krivoglaz M. A., Ryaboshapka K. P. and Barabash R. N., Theory of X-ray scattering by crystals containing dislocation walls, *Fiz. Met. Metalloved.* **30**, 1134–1145 (1970).
42. Kröner E., Zur plastischen verformung des vielkristalls, *Acta Metallurg.* **9**, 155–161 (1961).
43. Krug G. K., Sosulin Yu. A. and Fatuev V. A., *Design of Experiments in Problems of Identification and Extrapolation* (Nauka, Moscow, 1977) [in Russian].
44. Landau L. D. and Lifshitz E. M., *Course of Theoretical Physics*, Vol. 5: *Statistical Physics*, 3nd ed. (Nauka, Moscow, 1976; Pergamon, Oxford, 1980).
45. Landau L. D. and Lifshitz E. M., *Course of Theoretical Physics*, Vol. 7: *Theory of Elasticity*, 3nd ed. (Nauka, Moscow, 1965; Pergamon, Oxford, 1980).
46. Landau L. D. and Lifshitz E. M., *Course of Theoretical Physics*, Vol. 10: *Physical Kinetics* (Nauka, Moscow, 1979; Pergamon, Oxford, 1981).
47. Lekhnitsky S. G., *Theory of Elasticity of Anisotropic Body* (Gostekhizdat, Moscow–Leningrad, 1950) [in Russian].
48. Lifshitz I. M. and Rozentsveig L. N., On theory of elastic properties of polycrystals, *Zh. Exp. and Theor. Fiz.* **16**, 967–980 (1946).
49. *Mathematical Theory of Experiment Design*, Ed. C. M. Ermakov (Nauka, Moscow, 1983) [in Russian].
50. Miklyaev P. G. and Fridman Ya. B., *Anisotropy of the Mechanical Properties of Materials* (Metallurgiya, Moscow, 1986) [in Russian].
51. Morris R. R., Elastic constants of polycrystals, *Int. J. Eng. Sci.* **8**, 49–61 (1970).
52. Novikov I. I., *Theory of Thermal Treatment of Metals* (Metallurgiya, Moscow, 1986) [in Russian].
53. Nye J. F., *Physical Properties of Crystals, their Representation by Tensors and Matrices* (Mir, Moscow, 1967).
54. Olszak W. and Urbanowski W., The plastic potential and the generalized distortion energy in the theory of non-homogeneous anisotropic elastic-plastic bodies, *Archive Mech. Stos.* **8**, 671–694 (1956).
55. *Physical Metallurgy*, Ed. R. W. Cahn, Ch. XIII–XX (North-Holland Publishing Company, Amsterdam, 1965; Mir, Moscow, 1968).
56. Roe R.–J., Description of crystallite orientation in polycrystalline materials. III. General solution to pole figure inversion, *J. Appl. Phys.* **36**, 2024–2031 (1965).
57. Roe R.–J., Inversion of pole figures for materials having cubic crystal symmetry, *J. Appl. Phys.* **37**, 2069–2072 (1966).

58. Röpke G., *Statistische Mechanik für das Nichtgleichgewicht* (VEB Verlag, Berlin, 1987; Mir, Moscow, 1990).
59. Rozanov Yu. A., *Random Processes* (Nauka, Moscow, 1979) [in Russian].
60. Ruhl R. C. and Cohen M, Splat quenching of iron carbon alloys, *Trans. Metall. Soc. AJME* **245**, 241–253 (1969).
61. Ryaboshapka K. P., Possibilities of X-ray analysis of dislocation structures of deformed crystals, *Zavod. Lab.* **47** (5), 26–33 (1981).
62. Sachs L. *Statistische Schätzung* (Statistika, Moscow, 1976).
63. Satdarova F. F., X-ray scattering by deformed crystals, *Fiz. Met. Metalloved.* **49**, 467–480 (1980).
64. Satdarova F. F. and Kozlov D. A., Texture analysis by measurements of diffraction intensity in reciprocal space of polycrystal, *Fiz. Met. Metalloved.* **60**, 948–954 (1985).
65. Satdarova F. F., Kozlov D. A. and Blekhman B. N., Generalized parameters for dispersion of orientations of crystals in plane-deformed metals, *Fiz. Met. Metalloved.* **61**, 149–152 (1986).
66. Satdarova F. F., Analysis of the texture function, *Fiz. Met. Metalloved.* **69**, 204–207 (1990).
67. Satdarova F. F., Textural transformation upon recrystallization of sheet low-carbon steel, *Phys. Met. Metallography* **95**, 47–50 (2003).
68. Satdarova F. F., Dislocation structure of martensitic transformation in carbon steel, *Phys. Met. Metallography* **117**, 355–363 (2016).
69. Satdarova F. F. and Kiselev I. K., Analysis of diffraction spectra. I. On optimality and stability in the estimation of structure parameters, *Kristallografiya* **35**, 28–32 (1990).
70. Satdarova F. F. and Kiselev I. K., Analysis of diffraction spectra. II. On X-ray structural study of martensite of steel, *Kristallografiya* **35**, 33–37 (1990).
71. Satdarova F. F., Determination of the components of rhombic texture in metals, *Kristallografiya* **36**, 304–309 (1991).
72. Satdarova F. F., Analysis of a random system of dislocations in a deformed crystal, *Crystal. Rep.* **50**, 427–434 (2005).
73. Satdarova F. F., Identifiability of dislocation structures of strongly distorted crystals, *Crystal. Rep.* **51**, 72–80 (2006).
74. Satdarova F. F. and Kozlov D. A., Determination of the inhomogeneous dislocation density in a crystallographic texture, *Crystal. Rep.* **52**, 284–291 (2007).
75. Satdarova F. F., Distribution of dislocation density in deformed polycrystals, *Crystal. Rep.* **54**, 283–287 (2009).
76. Satdarova F. F., Plastic flow state and evolution of texture of metal sheet, *Izv. Ross. Akad. Nauk: Mekh. Tverd. Tela* № 6, 85–96 (2005).
77. Satdarova F. F., Plastic flow state and evolution of texture of metal sheet. II. Numerical analysis of deformation processes in a polycrystalline system, *Izv. Ross. Akad. Nauk: Mekh. Tverd. Tela* № 3, 135–143 (2006).

78. Satdarova F. F. and Kozlov D. A., Optimal design of experiment for measuring the texture function, *Zavod. Lab.* **48** (3), 44–48 (1982).
79. Satdarova F. F. and Kozlov D. A., Optimum estimation of harmonics of the physical profile of the diffraction line, *Zavod. Lab.* **48** (10), 53–56 (1982).
80. Satdarova F. F., Kozlov D. A. and Blekhman B. N., On the technique of quantitative measurements of texture, *Zavod. Lab.* **49** (3), 68–72 (1983).
81. Satdarova F. F., Kozlov D. A. and Kekalo A. I., Estimation of the effective elasticity coefficients of plane-deformed metals with a cubic crystal lattice from the measured harmonics of the texture function, *Zavod. Lab.* **51** (11), 53–56 (1985).
82. Satdarova F. F., Chernyshev A. V. and Yagubov S. A., Dialog system for texture analysis with optimal design of experiment, *Zavod. Lab.* **55** (1), 98–99 (1989).
83. Satdarova F. F., Critical analysis of texture estimation using simulation experiments, *Zavod. Lab.* **60** (3), 16–23 (1994).
84. Satdarova F. F., Modern texture analysis on the example of cold-rolled sheet of low-carbon steel, *Zavod. Lab. Diagn. Mater.* **68** (2), 21–25 (2002).
85. Satdarova F. F., On the application of quantitative texture analysis in technological studies in the making of metal sheet, *Zavod. Lab. Diagn. Mater.* **69** (7), 24–28 (2003).
86. Satdarova F. F. and Kozlov D. A., Automated system for analysis of the dislocation structure in polycrystals, *Zavod. Lab. Diagn. Mater.* **74** (9), 26–31 (2008).
87. Satdarova F. F., Analysis of diffraction measurements with adapting to background level, *Zavod. Lab. Diagn. Mater.* **75** (8), 35–38 (2009).
88. Satdarova F. F., Determination of the physical profile of the diffraction line by the pseudo-maximum likelihood method, *Zavod. Lab. Diagn. Mater.* **78** (5), 73–77 (2012).
89. Savyolova T. I. and Ivanova T. M., Methods for reconstructing the distribution function of orientations by pole figures, *Zavod. Lab. Diagn. Mater.* **74** (7), 25–33 (2008).
90. Shermergor T. D., *Theory of Elasticity of Microinhomogeneous Media* (Nauka, Moscow, 1977) [in Russian].
91. Spector E. N., Gorelik S. S. and Rakhshtadt A. G., *Izv. VUZ: Tsvetnaya metallurgiya* №.4, 132–134 (1971).
92. *Tables of Integral Transforms* (*Bateman Manuscript Project*) Director A. Erdélyi, Vol. 2 (McGraw-Hill, New York, 1954; Nauka, Moscow, 1970).
93. Tarantsev A. A., On the relation between interval analysis and probability theory, *Zavod. Lab. Diagn. Mater.* **70** (3), 61–65 (2004).
94. Teodosiu C., *Elastic Models of Crystal Defects* (Springer-Verlag, Berlin, Heidelberg, New York, 1982; Mir, Moscow, 1985).
95. Tikhonov A. N. and Arsenin V. Ya., *Methods for Solving Ill-Posed Problems* (Nauka, Moscow, 1979).

96. Trushkovski V., Vezhbinski S. and Kloch Ya., Quantitative analysis of the plastic anisotropy of polycrystalline copper, *Fiz. Met. Metalloved.* **66**, 178–183 (1988).
97. Ungar T. and Borbely A., The effect of dislocation contrast on X-ray line broadening; a new approach to line profile analysis, *Appl. Phys. Letters* **69**, 3173–3175 (1996).
98. Utevsky L. M., *Diffraction Electron Microscopy in Metallurgy* (Metallurgiya, Moscow, 1973) [in Russian].
99. Viglin A. S., A quantitative measure of texture of a polycrystalline material. Texture function, *Fiz. Tverd. Tela* **2**, 2463–2476 (1960).
100. Vilenkin N. Ya., *Special Functions and Theory of Groups Representations* (Nauka, Moscow, 1965) [in Russian].
101. Vlasov A. A., *Statistical Distribution Functions* (Nauka, Moscow, 1966) [in Russian].
102. Warren B., X-ray analysis of deformed metals, *Prog. Metal. Phys.* **8**, 147 (1959).
103. Watson G. S., The Statistics of Orientation Data, *J. Geology* **74** Pt. 2, 786–797 (1966).
104. Weiss M. and Montagnat M., Long-range spatial correlations and scaling in dislocation and slip patterns, *Philosophical Magazine* **87**, 1161–1177 (2007).
105. Yanossy L., *Theory and Practice of the Evaluation of Measurements* (Oxford Univ., 1965; Mir, Moscow, 1968).
106. Young R. A., Gerdes R. J. and Wilson A. J. C., Propagation of some systematic errors in X-ray line profile analysis, *Acta Cryst.* **22**, 155–162 (1967).
107. Zasimchuk E. A. and Isaichev V. I., Mechanical instability of a fragmented structure in terms of nonlinear thermodynamics, *Doc. Akad. Nauk SSSR* **302**, 1101–1104 (1988).
108. Zolotarevsky I. Yu. and Rybin V. V., Deformation of fragmenting polycrystals and texture formation, *Fiz. Met. Metalloved.* **59**, 440–449 (1985).
109. Zolotarevsky I. Yu., Rybin V. V. and Zhukovsky I. M., Theory of deformation textures of fragmenting metals, *Fiz. Met. Metalloved.* **67**, 221–232 (1989).
110. Zubarev D. N., *Nonequilibrium Statistical Thermodynamics* (Nauka, Moscow, 1971) [in Russian].

TABLE OF CONTENTS